Earthquake Resistant Design
of Buildings

Earthquake Resistant Design of Buildings

Mehmet E. Uz

Department of Civil Engineering, Adnan Menderes University
Aydin Turkey

Former, Associate Research Fellow, ARC
Research Hub for Australian Steel Manufacturing
Wollongong, NSW
Australia

and

Muhammad N.S. Hadi

School of Civil, Mining and Environmental Engineering
University of Wollongong, Wollongong, NSW
Australia

CRC Press
Taylor & Francis Group
Boca Raton London New York

CRC Press is an imprint of the
Taylor & Francis Group, an **informa** business

A SCIENCE PUBLISHERS BOOK

CRC Press
Taylor & Francis Group
6000 Broken Sound Parkway NW, Suite 300
Boca Raton, FL 33487-2742

First issued in paperback 2020

© 2018 by Taylor & Francis Group, LLC
CRC Press is an imprint of Taylor & Francis Group, an Informa business

No claim to original U.S. Government works

ISBN-13: 978-0-8153-9172-2 (hbk)
ISBN-13: 978-0-367-78183-5 (pbk)

**Visit the Taylor & Francis Web site at
http://www.taylorandfrancis.com**

**and the CRC Press Web site at
http://www.crcpress.com**

Features

- Introduces the main seismic concepts and the performance and response of structures to earthquakes
- Reviews the effect of the use of dampers on dynamic characteristics of building models with respect to shear wave velocity on the Soil-Structure Interaction (SSI) systems, the available modal analysis methods for SSI systems and their application to Multi Degree of Freedom (MDOF) modal response history analyses
- Presents a formulation of MDOF adjacent buildings used for analytical study of the book
- Offers the optimum design, using genetic algorithm, NSGA-II and Pareto optimal solution, having discussed control algorithms

Preface

While introducing important concepts in the study of earthquakes related to retrofitting of structures to be made earthquake resistant. The book investigates the pounding effects on base-isolated buildings, the soil-structure-interaction effects on adjacent buildings due to the impact, the seismic protection of adjacent buildings and the mitigation of earthquake-induced vibrations of two adjacent structures. These concepts call for a new understanding of controlled systems with passive-active dampers and semi-active dampers. The passive control strategy of coupled buildings is investigated for seismic protection in comparison to active and semi-active control strategies.

Acknowledgement

First Author: First and foremost, I would like to express my profound gratitude to the Rector of Adnan Menderes University, Prof. Dr. Cavit Bircan, Aydin, Turkey for his invaluable direction, remarkable mentorship and for accepting me as a Lecturer at Adnan Menderes University and also for his help with all administrative matters in Ankara. His constant support and encouragement and understanding of what makes us human are gratefully acknowledged. I would like to especially acknowledge Mrs. Hatice Bircan for helping to keep us on track and providing the much needed support and humour along the way. I am grateful to Mr. Talha Bircan for immeasurable enthusiasm and sharing his brilliant thoughts and ideas. Bircan family guided me to a wider world, making my life enjoyable and meaningful and I am grateful to call all—my big family in Aydin.

My heartiest thanks to Dr. Hayfa Hadi: I am thankful beyond words for her support and undying love and I am grateful to call her my mother. I also would like to thank my fellow researchers, Dr. Lip Teh at the University of Wollongong, Wollongong, for his continuous support and academic insights, which kept me motivated to work harder. I am indebted to the Turkish Government for providing financial assistance through my education that enabled me to pursue my research for this book. I would like to thank the Department of Civil, Faculty of Engineering, University of Adnan Menderes University, for providing all necessary facilities and suitable conditions for my research. I am deeply grateful to some families in Australia, Mrs. Ida Teh, Mr. Ahmed Hadi, Mrs. Reema Hadi and my best little friends Selim, Efna, Zach, Suleyman, Lewis and Emir.

Finally, I would like to express my deepest gratitude to my family, whose constant love and prayer became an active strength during the completion of this research. They have always believed that I could achieve anything I tried hard enough to achieve. I am sincerely grateful to my wife Chi-Jung (Chloe) Lai Uz for her understanding and patience during my research for this book, without her my life would not have been this pleasant.

Contents

Contents

1

Effects of Earthquake on Structures

1.1 Preview

In the context of structural design, optimisation is essential to create structural systems that maximise the levels of safety and economise on the limited resources available in computerised structural design methods. Residential buildings, which are located in seismically-active regions, are often built close to each other due to the economics of land use or architectural reasons. Pounding refers to a building hitting an adjacent building due to lateral loading, for example, an earthquake, which is one of the main causes of severe building damages. Non-structural damage means movement across seperation joints between adjacent buildings. Due to the closeness of structures, the pounding problem is especially widespread and dangerous since a maximum land use is required. The simplest way to reduce or avoid pounding is to provide adequate separation distance between the buildings. The required separation distance is not always easy to provide. Thus, a minimum separation distance is desired between adjacent buildings. This book describes a range of pounding reduction devices, and a comprehensive literature review of topics that form the background. The problem is complicated by the fact that adjacent buildings with different owners are built at different times with different building code specifications. These buildings can be designed with different dynamic characteristics. Seismic pounding between adjacent buildings with different dynamic characteristics can occur because of the lack of an essential gap between the buildings. Further, the book shows the effect of the use of dampers on the dynamic characteristics of the building models, with respect to shear wave velocity on the soil-structure interaction (SSI) systems, the available modal analysis methods for SSI systems and their application to Multi Degree Of Freedom (MDOF) modal response history analyses. The main differences and advantages are pointed out in addition to the optimal design of passive, active and semiactive devices in structural control optimisation and design approaches.

1.2 Objectives and Limitations

Investigation of both the pounding response, including soil-structure interaction (SSI) and pounding of seismically isolated buildings, without considering the presence of adjacent buildings is very limited. By considering the recent developments in the structural control methods, the concept of control system for adjacent buildings plays an important role. Thus, this book develops a methodological framework that is applicable to various types of structural engineering projects. These projects can be a design and/or retrofit of structures with some control devices in the scope of structural optimisation problems. In addition, according to a certain performance index, the designed control system needs to be optimum in some sense. Optimum dampers, parameters of passive control devices, command voltage of semi-active dampers and stable controllers of active control systems are obtained according to the chosen performance index.

Some assumptions have been adopted during the calculations for highlighting the important features of damper devices connected to neighbouring buildings and investigating the effect of soil-structure interaction on the response of fixed-base isolated buildings. The rigid diaphragm of the floor is assumed. A linear multi-degree of freedom system where the mass of each level is lumped in the floors is provided for each building. The stiffness is provided in terms of columns. Moreover, the friction coefficients between the sliding surfaces are taken as constant while the base isolated buildings are investigated. Spatial difference of the ground motion can be neglected because the total plan dimensions of the buildings in the excitation direction are not large. Pounding forces are calculated by the Coulomb friction model. Hence, in the y planer and z vertical directions, the coefficient of friction and coefficient of restitution for energy dissipation are assumed to be constant during the impact.

The SSI forces are modelled in the form of frequency-independent soil springs and dashpots with modelling the coupled buildings resting on the surface of an elastic half-space. An investigation is carried out in three parts for the specific objectives of this book. Firstly, the investigation is done to analyse the earthquake-induced pounding between two insufficiently separated buildings, considering the inelastic behaviour of the structures' response. Secondly, the seismic response history analysis of multi-storey inelastic adjacent buildings of different sizes with SSI systems during the impact is investigated. After the first two parts, the specific objectives of this book continue to focus on the reduction of displacement, acceleration and shear force responses of adjacent buildings, using supplemental damping devices. Further, the optimal placement of the damper devices, instead of placing them on all the floors in order to minimise the cost of the dampers, is investigated. Various earthquake records are used to examine the seismic response of two buildings under different ground motions. A formulation of the equation of motion for two different buildings is presented for each numerical example used in this book. The resulting systems of second-order constant coefficient equations are reformulated as a system of first-order ordinary differential equations and solved, using the ordinary differential equation solver of MATLAB. The coupled multi-degrees of freedom modal differential equations of motion for two-way asymmetric shear buildings are derived and solved, using a step-by-step solution by the fourth-order Runge-Kutta method with and without any impact in this book.

A conventional optimisation approach is combined with multiple objective functions into a single-objective function by use of arbitrary weights. The three objectives of this book are to minimise the number of MR dampers for economical benefits, the H_∞ norm transfer function from external disturbance to the regulated output and the peak displacement or storey drift responses non-dimensionalised by related responses of the uncontrolled system. Numerical results of adjacent buildings controlled with MR dampers and the corresponding uncontrolled results are examined and compared with non-linear control algorithms. The optimal design of semi-active dampers placed between adjacent buildings is investigated in two different parts. The binary coded GA automatically employs and optimises the controllers used in this book in accordance with the fitness function that reflects the multi objective. In the first part, an adaptive method for design of a Fuzzy Logic Control (FLC) system for protecting adjacent buildings under dynamic hazards using MR dampers is proposed in single GA. Design of the Genetic Adaptive Fuzzy (GAF) controller is conducted in the first part of this book. Minimisations of the peak interstorey drift and displacement related to ground responses are the two objectives. A global optimisation method which is a modification of binary coded genetic algorithm adopted by Arfiadi and Hadi (2000, 2001) is used. Binary coded GA is used to derive an adaptive method for selection of fuzzy rules of the FLC system. The fuzzy correlation between the inputs (structural responses) and the outputs (command voltages) of the controller is provided by adding, changing and deleting the rules of the FLC system. Inputs are taken as top-floor displacements of both the buildings. Nevertheless, the multi objectives combined with single genetic algorithm and NSGA-II are also directly used as controllers to determine the vectors of both the number of dampers and the command voltage for each floor.

The responses of the adjacent buildings are compared with the corresponding uncontrolled individual buildings. Further, three proposed controllers are compared under the other controllers. The other controllers are passive (on zero or maximum command voltage), and semi-active controllers based on LQR and H_2/LQG (command voltage governed by the clipped optimal control law). In the last part, this book concludes that the main objectives of the optimal damper design are not only to reduce the seismic responses of the adjacent structures but also to save the total cost of the damper system. Therefore, the peak inter-storey drift response and the total number of non-linear dampers constitute the objective functions of the optimisation problem. The influence of damper location and the regulated outputs to be minimised on the control performance are studied. In short, the main objectives of the book are as follows:

- Investigation of seismic response analysis for the behaviour of inelastic adjacent buildings for comparison of fixed-base and base-isolated buildings modals under two-directional ground motions considering the impact effects, using a modified non-linear visco-elastic model

- Identification of the key structural parameters influencing the pounding response of buildings and investigation of the effect of varying one parameter under earthquake excitations
- Development of an efficient approximate model that can fully describe the seismic response history analysis of elastic and inelastic multi-storey adjacent buildings of different dimensions with SSI effect during the impact due to earthquake excitations based on conventional modal response history analysis
- Presentation of total response histories of the various SSI systems, using MDOF modal equations of motion and comparing with the equation of motion for the whole SSI system obtained by the direct integration method
- Improvement of application of the structural control concepts for linking coupled buildings by Fluid Viscous Dampers (FVD) with or without the actuator to obtain the parameters or controller gains to achieve the best structural performance
- Identification of optimal placement of the damper devices instead of placing them on all the floors in order to minimise the cost of dampers using genetic algorithm
- Development of application of new structural control concepts for adjacent buildings, using MR dampers between the adjacent buildings with different controllers for optimising the parameters in single genetic algorithm and Non-Sorting Genetic Algorithm-II (NSGA-II).

1.3 Review of Literature

The literature review is carried out in four main parts. The first two parts concentrate on the effect of pounding on adjacent buildings with and without base isolation, considering the historical earthquakes. In the last two parts, several control strategies that were proposed by researchers are discussed. Studies in conjunction with the Soil-Structure Interactions (SSI) systems, using either frequency-dependent or frequency-independent for soil springs and dashpots, are investigated in this section.

1.3.1 Earthquakes

Earthquakes have always been important to structural engineering in terms of the dynamic response of building structures. The damage reports after strong earthquakes show that the effect of structural pounding is often one of the reasons for the damage. Examples of serious seismic hazards due to pounding were reported during the Alaska (1964), San Fernando (1971), Mexico City (1985), Loma Prieta (1989) and Kobe (1995) earthquakes. Recent damaging earthquakes, such as at Northridge (1994), Kobe (1995), Turkey (1999) and Taiwan (1999) showed that the vibration of structures was so severe that it led to the collapse of buildings (Kasai et al. 1996). In order to understand and prevent poundings, some measures at pounding locations (Anagnostopoulos and Spiliopoulos 1992, Jankowski et al. 1998) can be minimum separation distance between structures (Kasai et al. 1996, Penzien 1997, Lopez-Garcia and Soong 2009) and using some damper devices (Westermo 1989, Luco and Debarros 1998, Yang et al. 2003, Hadi and Uz 2009, Uz and Hadi 2009). In order to mitigate the structural systems response, control devices and schemes are being considered and commonly used in modern structural design. In order to reduce the seismic response of buildings, control devices have been utilised as supplement damping strategies.

Building complexes have often impacted each other under earthquake-induced strong ground motion. A sudden movement of the ground caused by an earthquake can be transferred to the structure through the foundation. Chopra (1995) determined a time history of the ground acceleration usually defined by ground motion during an earthquake. Nowadays, various types of base isolation systems in coupled buildings are being proposed by several researchers in order to reduce the response of each structure during earthquake excitations. One approach is to modify the foundation of a superstructure by inserting a layer of material that has very low lateral stiffness, thus reducing the natural period of vibration of the structure. This approach is common in retrofitting structures by using a discrete number of friction bearings located between the foundation and the superstructure. Moreover, the existing space between buildings may not be enough to avoid pounding if either historic restoration or seismic rehabilitation for existing fixed-base buildings is done with the use of base isolation systems.

1.3.2 Seismic Behaviour of Adjacent Buildings

Based on the above facts, there is a need to study the effect of base isolation on pounding of buildings as well as on pounding of these base-isolated buildings. The characteristics of pounding were evaluated on existing building configurations to provide guidance for future building designs by Maison and Kasai (1992). The installation of friction bearing allows the structure to slide on its foundation during ground movement. Hence, Agarwal et al. (2007) studied the use of friction bearings to reduce the base shear force transferred to the structure. Moreover, the use of friction bearings reduces the displacement of floors with respect to the base and the internal forces created in the structure. They can also be designed to permit sliding of the structure during very strong earthquakes. Hence, an investigation of the effect of base isolation can become a need to further explore the pounding of buildings. The differences in geometric and material properties of adjacent buildings can lead to significant differences in their response to the external forces. Spatial variation of the ground movement in case of an earthquake can also cause the adjacent structures to respond differently. As a result, adjacent buildings may vibrate quite differently during an earthquake and may impact each other at various times during their movement. Due to the huge mass of buildings, the momentum of vibrating structures increases, causing a lot of local damage during an impact.

In addition, large impact forces can significantly change the response behaviour during an impact or building pounding. The pounding phenomenon which occurs in multi-frame bridges, where the decks of the bridge impact each other during an earthquake, was investigated by DesRoches and Muthukumar (2002). The Mexico City earthquake (1985) is a good example of the degree of damage the pounding causes. Many cases of structural damage owing to mutual pounding between adjacent buildings during a major earthquake have been reported over the last two decades. Thereby, Lin and Weng (2001) determined a numerical simulation approach to estimate the probability of seismic pounding on adjacent buildings, using provisions of the 1997 Uniform Building Code. In order to determine the required distance between buildings, the International Building Code (2000) specifies the distance to be maintained between adjacent buildings as the square root of the sum of squares of their individual displacements. Kasai et al. (1996) used the spectral difference (SPD) method to calculate adequate building separation distance. Jankowski (2008) found that simulation errors in pounding force histories were equal to 12.7 per cent for the linear viscoelastic model, 33.5 per cent for the non-linear elastic model and 11.6 per cent for the non-linear viscoelastic model.

Another method to study the pounding problem is the 'equivalent static force' method in which equivalent static horizontal forces are applied to the building to simulate an earthquake. Matsagar and Jangid (2003) studied the seismic response of multi-storey building supported on different base isolation systems during an impact with adjacent structures. The step-by-step iteration method in the Newmark formulation for the coupled differential equations of motion in the isolated system is used and solved in an incremental form. With the difference of main system characteristics, such as the distance of gap, stiffness of impact element, superstructure flexibility and number of storeys of base-isolated building, the impact response of isolated building is studied. It is concluded that the effects of impact are found to be severe in a system with flexible superstructure, increased number of storeys and bigger stiffness in the adjacent structure.

1.3.3 Structural Control Optimisation

In the design of structures using control systems, the objective of this section is to provide an overview of research studies in the development of optimisation methods. Use of dampers for structures is reviewed under three approaches in this section. They are analytic formulations for design of dampers with parametric studies, optimal control theory and genetic algorithm approaches. Despite high-rise buildings being constructed in close proximity to each other, various methodologies for analytical formulations of interconnecting adjacent buildings have been examined for seismic hazard mitigation. Kim et al. (2006) investigated the effect of installing Visco-Elastic Dampers (VEDs) in places, such as building–sky-bridge connections, to reduce earthquake-induced structural responses. In order to investigate the effectiveness of the proposed scheme, a parametric study was undertaken, using single-degree-of-freedom systems connected by VEDs and subjected to white noise and earthquake ground excitations. Dynamic analyses were carried out with 5-storey and 25-storey rigid frames. According to their analysis results, the use of VEDs in sky-bridges can be effective in reducing earthquake-induced responses if they are designed in such a way that the natural frequencies become quite different. This difference can be achieved by connecting with VEDs having different structural systems. Qi and Chang (1995) described the implementation of viscous dampers possessing several inherent and significant advantages, including linear viscous behaviour; insensitivity to stroke and output force; easy installation; almost-free maintenance and reliability and longevity.

Richard et al. (2006) mentioned coupled building control as an effective means of protection in flexible building structures. They studied the effects of building configuration and the location of connector on the overall system performance. They also examined the efficacy of passive and active coupled building control for flexible adjacent buildings. Zhu and Xu (2005) determined the analytical formulae for determining optimum parameters of Maxwell model-defined dampers to connect two coupled structures, using the principle of minimising the average vibration energy of both the primary structure and the two adjacent structures under white-noise ground excitation. A dynamic analysis shows that the damper of optimum parameters can significantly reduce the dynamic responses of the adjacent structures under white-noise ground excitation. Hadi and Uz (2009) investigated the importance of viscous fluid dampers for improving the dynamic behaviour of adjacent buildings by connecting them with fluid viscous dampers.

Kageyama et al. (1994) proposed a method to reduce the seismic response of a double-frame building by connecting the inner and outer structures with dampers. Iwanami et al. (1996) investigated the optimum damping and stiffness values of the connecting damper by assuming each of the linked structures as a single degree of freedom system. Luco and Debarros (1998) examined the optimal distribution of the connecting dampers by modelling the neighbouring floors. Sugino et al. (1999) and Arfiadi and Hadi (2001) calculated the optimal parameters of the connecting dampers in conjunction with the genetic algorithm. Ni et al. (2001) analysed the seismic response of two adjacent buildings connected with non-linear hysteretic damping devices under different earthquake excitations and demonstrated that non-linear hysteretic dampers are effective even if they are placed on a few floor levels. A parametric study also shows that optimum damper parameters and numbers are significantly important parameters in order to minimise the random seismic response. They also investigated the effect of variation of placement and the number of dampers on the seismic response of a structure.

Moreover, using dampers with optimal parameters to link the adjacent buildings can increase the modal damping ratios. Thus, optimal parameters of the passive element, such as damping and stiffness under different earthquake excitations, can influence the structural parameters of the system (Yang et al. 2003). Ying et al. (2003) investigated a stochastic optimal coupling-control method for adjacent building structures. In a reduced-order model for control analysis, the coupled structures with control devices under random seismic excitations are modelled in their study. With structural energy control, both the seismic response mitigation for adjacent building structures and the dimension of optimal control problem are reduced. For a random response, non-linear controlled buildings and uncontrolled buildings are predicted by using the stochastic averaging method to evaluate the control efficacy. They also conducted a numerical study to demonstrate the response reduction capacity of the proposed stochastic optimal coupling-control method for adjacent buildings.

Mitigation of the seismic response of adjacent structures connected with active control devices has been investigated by Seto and Mitsuta (1992), Luco and Wong (1994), Yamada et al. (1994), Arfiadi and Hadi (2000) and Kurihara et al. (1997). Various control strategies were investigated by a number of researchers and full-scale applications are beginning to appear. Spencer et al. (1998) studied a scale model of a three-storey building with active mass driver and active tendon system, using various active control algorithms. In the optimal control theory approach, optimising the use of dampers to mitigate seismic damage has hitherto not been investigated in spite of enhancing structural control concepts in structural vibration control through the application of optimisation. Luco and De Barros (1998) investigated the optimal damping values for distribution of passive dampers interconnecting two adjacent structures. In general, analytical and experimental studies investigated the dynamic responses of the structures before and after installing a damping device to understand their effectiveness. However, very few studies have been done with regard to the effect of non-uniform distribution of dampers (Yang et al. 2003, Bhaskararao and Jangid 2006a, Ok et al. 2008, Bharti et al. 2010). None of these studies show a clear comparison in order to indicate the quality of their proposed arrangement/solution. For example, Bhaskararao and Jangid (2006a) proposed a parametric study to investigate the optimum slip force of the dampers in the responses of two adjacent structures. The authors also showed that the response reduction is associated with optimum placement of damper.

Yang et al. (2003) showed that in order to minimise the loss in a performance, the number of dampers can be decreased. The authors confirmed that it is not necessary to equip every single floor with a viscous damper based on their trial-and-error solution, but no solution for the optimal arrangement was provided. A similar study for semi-active magnetorheological (MR) dampers was conducted by Bharti et al. (2010) who proposed that the placement of damper is not neccessary for every single floor. They also confirmed the results obtained by Ok et al. (2008) in a performance for adjacent buildings equipped with MR dampers by use of genetic algorithms. In the study by Ok et al. (2008), an optimal design method of non-linear hysteretic damper as an example of passive type magnetorheological (MR) damper was conducted based on multi-objective genetic algorithm and nonlinear random vibration analysis with the stochastic linearisation method. It was found that the building parameter and damper parameters influence the response of connected buildings (Bharti and Shrimali 2007, Dumne and Shrimali 2007). In an attempt to devise a clear method to provide the optimal arrangement, Bigdeli et al. (2012) introduced optimisation algorithms to find the optimal configuration for a given number of dampers. For the purpose of comparison, the authors also used the highest relative velocity heuristic approach based on the work of Uz (2009). Hadi and Uz (2009) and Uz and Hadi (2009) proposed placement of fluid viscous dampers in floors where the maximum relative velocity occurs. This work was repeated by Patel and Jangid (2009).

During the last two decades, a procedure was developed in order to analyse three-dimensional buildings utilising passive and active control devices, in which two types of active control devices—an active tuned mass damper and an active bracing system—were taken into account by Arfiadi and Hadi (2000). The passive parameters of the dampers as well as the controller gain were then optimised, using a genetic algorithm based on the LQR, H_2, H_∞ and L_1 norms of the objective function. Although a passive control technique is adopted due to its simplicity, semi-active and active control systems receive considerable attention now. Arfiadi and Hadi (2001) improved a simple optimisation procedure with the help of GAs to design the control force. They used a static output feedback controller utilising the measurement output. In this case, the control force is obtained by multiplying the measurement with the gain matrix (Arfiadi and Hadi 2001).

In the genetic algorithm approach (GA), which was initially developed by Holland (1975), was used as an optimisation tool in designing control systems (Hadi and Arfiadi 1998, Arfiadi 2000, Arfiadi and Hadi 2000, 2001, 2006, Bigdeli et al. 2012, Sun et al. 2013). Control algorithms developed for passive, semi-active and active control have been directly useful for developing other recent control strategies. The most common optimal control algorithms, such as Linear Quadratic Regulator (LQR), H_2/LQG (Linear Quadratic Gaussian), H_2, H_∞ and fuzzy control can be chosen by combining the GAs. The objectives of minimising peak inter-storey drift and the total floor acceleration response of the structure were combined into a single function to find the optimal distribution and size of PED devices under earthquake excitation in the seismic design of shear frame structure by Dargush and Sant (2005). Ahlawat and Ramaswamy (2003) proposed an optimum design of dampers, using a multi-objective version of the GA. Application of the Multi Objective Genetic Algorithm (MOGA) in determining the design and retrofitting of non-linear or linear shear frame structures with passive energy devices was recently proposed by Lavan and Dargush (2009). In their study, the maximum inter-story drift and the maximum total floor acceleration were simultaneously minimised by using the MOGA. The optimal size, location and type of dampers were determined to achieve their objective in the study. However, this method has been applied only to a specific optimisation problem. Therefore, this study presents different types of single and multi-objective optimisation problems by using global and local objective function measures and applied for optimal design of passively, actively and semi-actively-controlled adjacent buildings. For obtaining the best results in the reduction of structures, combined application of GAs and fuzzy logic was proposed to design and optimise the different parameters of active dampers by Pourzeynali et al. (2007).

The fuzzy logic control (FLC) theory has attracted the attention of engineers during the last few years (Symans and Kelly 1999, Yan and Zhou 2006), but there are some drawbacks in FLC systems. The fuzzy sets and rules that require a full understanding of the system dynamics must be correctly pre-determined for the system to function properly. Furthermore, in order to mitigate the responses of seismically-subjected civil engineering structures, multiple MR dampers distributed between the adjacent buildings should be used (Yan and Zhou 2006). Zhou et al. (2003) successfully applied an adaptive fuzzy control strategy for control of linear and non-linear structures. The authors found that the adaptive feature of a fuzzy controller has various advantages in the control of a building, including an MR damper system. GA optimisation of FLCs related to structural control applications was recently investigated by Kim and Roschke (2006). They proposed minimisation of four structural response objectives utilising a non-dominated sorting genetic algorithm (NSGA-II) in each of these application control of a hybrid base isolation system by MR dampers.

Another trend in the development of FLC system is to combine genetic algorithms (GA) as an optimisation tool in designing control systems (Hadi and Arfiadi 1998, Arfiadi 2000, Arfiadi and Hadi 2000, 2001, Kim and Ghaboussi 2001, Kim et al. 2006, Bigdeli et al. 2012). Optimising the dampers to mitigate seismic damage to adjacent buildings has hitherto not been investigated in spite of enhanced structural control concepts in structural vibration control through the application of optimisation in an integrated GA-FLC. Ahlawat and Ramaswamy (2003) proposed an optimum design for dampers using a multi-objective version of the GA. Arfiadi and Hadi (2001) improved a simple optimisation procedure with the help of GAs to design the control force. They used a static output feedback controller utilising the measurement output. For obtaining the best results in the reduction of structures, combined application of GAs and FLC was proposed to design and optimise the different parameters of active dampers by Pourzeynali et al. (2007).

1.3.4 Soil-Structure Interaction

Interactions between inadequately separated buildings, soil-structure or bridge segments was repeatedly observed during severe ground motion. An accurate structural model has to be built to effectively represent the dynamic characteristics of real structures for investigating the response of structural systems under severe earthquake excitations (Wu et al. 2001). An efficient methodology was given in detail to systematically evaluate the dynamic responses of irregular buildings with the consideration of soil-structure interaction in the study of Wu et al. (2001) (see Fig. 1.1). Realistic structural models by incorporating the modelling complexities, such as pounding-involved response (Jankowski 2008, Hadi and Uz 2010b), soil-structure interaction (SSI) (Gupta and Trifunac 1991, Sivakumaran et al. 1992) and torsional coupling (Hejal and Chopra 1989) have recently been developed in various analytical approaches based on the progress made in structural dynamics and earthquake engineering. It is also recognised that SSI and torsional coupling may result in substantial damage or even lead to a structure's collapse during earthquakes owing to the fact that these interactions can drastically alter the seismic response of structures. In the past two decades, research studies elucidated that the reduction of natural

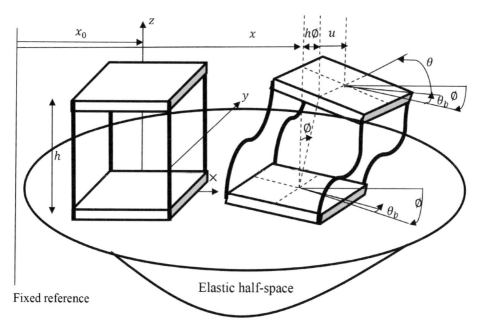

Fig. 1.1 Illustration of an asymmetric building with the soil-interaction model.

frequency due to soil flexibility, the partial dissipation of the vibration energy of a structure through wave radiation into the soil and the modification of actual foundation motion from the free field ground motion can be observed on foundation structure systems (Gupta and Trifunac 1991, Sivakumaran and Balendra 1994, Wu et al. 2001). However, the common practice usually does not account for the effects of earthquake-induced SSI on the seismic behaviour of adjacent buildings of different dimensions. The effect of SSI on structure response had not been seriously taken into account until the beginning of nuclear plant constructions. The earthquake response analysis of linear symmetric buildings with SSI has been investigated well enough (Novak and Hifnawy 1983a, Seto 1994, Spyrakos et al. 2009). In these studies, multistorey buildings resting on the surface of an elastic half space were modelled as a planar building for the interaction system.

Many researchers have investigated the issue of seismic analysis of two-way asymmetric buildings with soil-structure interaction under two-directional ground motions (Balendra et al. 1982, Balendra 1983, Balendra et al. 1983, Novak and Hifnawy 1983b, Sivakumaran and Balendra 1994). In these studies, to deal with non-proportional damping of the SSI systems, equivalent modal damping was calculated to facilitate the modal response history analysis. The equivalent modal damping was estimated by either quantifying the dissipated energy in the soil (Novak and Hifnawy 1983a, Novak and Hifnawy 1983b) or matching the approximation approaches normal mode solution with rigorous solution of a certain structural location (Balendra et al. 1982). For engineering applications without the need for calculating the complicated equivalent modal damping, a simple and real valued modal response history analysis has been developed (Tsai et al. 1975, Jui-Liang et al. 2009). In these studies, the modal response histories of a single building resting on the surface of an elastic half space were investigated by solving the multi-degree of freedom modal equations of motion, using the step-by-step integration method.

The aim of this book is to conduct a comparative research study to investigate the SSI effect on adjacent buildings, considering the effect of pounding. Based on the fourth order Runge-Kutta method, a MATLAB program was developed to solve the equations of motion for SSI systems of coupled buildings subjected to pounding effects under excitations of earthquake ground acceleration.

1.4 Classification of Seismic Isolation Devices

Figure 1.2 shows the classification of damper devices in the mitigation of seismic response of buildings. Adjacent high/medium-rise buildings with control devices to mitigate the seismic or wind response have been an active research subject in recent years. The aim of passive response-control systems, in which a kind of energy dissipation or damping mechanism is installed, is to mitigate the effects of earthquakes and/or wind induced vibration of buildings for improving and enhancing the performance of individual buildings (Kitagawa and Midorikawa 1998). Many types of energy dissipation or damping devices have been developed, such as a hysteretic damper (HD), a friction damper (FD), a tuned liquid (sloshing) damper (TLD), a tuned mass damper (TMD), an oil damper (OD), a fluid viscous damper (FVD) and a visco-elastic damper (VED) as shown in Fig. 1.2.

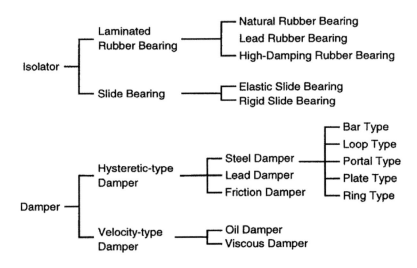

Fig. 1.2 Classification of seismic isolation devices.

The methods mentioned above can be classified as structural devices used to control the seismic movements of structures, where the mechanisms of structural response can be activated by motion of the structure itself or by motion of the control forces applying devices. The mechanisms of structural response are enhanced by using improved control devices in buildings.

1.5 Efficiency of Control Devices

Nowadays, the use of control devices in buildings has become inevitable to mitigate the structural vibration. Moreover, for improving the dynamic behaviour of adjacent buildings, control devices are connected between adjacent buildings as active, semi-active and passive devices. In order to ensure maximum safety of structures, today's modern concepts highlight the use of added damping, base isolation and mass damper as alternative methods for mitigating the total displacement and inter-storey drift of the structure. The concept of linking adjacent buildings using the above mentioned dampers has thus been proposed to improve their both seismic-resistant and wind-resistant performances. Passive devices are quite simple to design and build. However, the performance of passive control is sometimes limited. Therefore, for achieving a relative performance of passive control devices, optimum damper properties are implemented to protect against one particular dynamic loading. Active control strategies are generally effective, but they have some disadvantages while needing large amounts of power in action. Further, they may result in instabilities in controlled structures. Semi-active devices have been shown to possess the advantages of active control devices without requiring the associated large power sources. If power fails, they behave like a passive device. For these reasons, they have a promising future in structural control. A design system based on a performance function is utilised to obtain the damper parameters resulting in the best overall system response. The control parameters of devices, such as the damping capacity, number and location of a device in each of the adjacent buildings, have a significant effect on response mitigation. Further, these control parameters are important to achieve the desired design objectives while satisfying the constraints. In the context of structural design, optimisation is used to describe the progress of new structural systems in a manner that maximises safety and minimises the cost while satisfying certain constraints. Optimum design of structures with installed control devices was conducted for not only defining size, type and location of devices but also in choosing the cost function to be minimised.

1.6 Lumped Mass Systems

The response of older buildings adjacent to the site also needs to be considered. Several models of closely spaced adjacent buildings have been developed. These models can be categorised as lumped mass systems, which, being the most basic idealisation of a structure and relatively straightforward to analyse, are most popular. A three-dimensional model of

MDOF system using finite element models was developed by Papadrakakis et al. (1996). They used the Lagrange Multiplier Method to study the response of two or more adjacent buildings located in a series or in a orthogonal configuration with respect to one another.

As shown in Fig. 1.3, the adjacent buildings have been modelled at two degree of freedom (2-DOF) systems with lumped masses m_{11}, m_{12}, m_{21} and m_{22}. The stiffnesses of the two buildings are k_{11}, k_{12}, k_{21} and k_{22} and their linear viscous constants are c_{11}, c_{12}, c_{21} and c_{22} respectively. Two buildings have been modelled by introducing a spring and a linear viscous dashpot between the adjacent buildings. The stiffness of the spring between the first floors in adjacent buildings is $k_{11,21}$ and that between the second floors is $k_{12,22}$. The corresponding viscous constants are $c_{11,21}$ and $c_{12,22}$.

Dynamic equations for adjacent buildings connected by a damper in two degrees of freedom systems can be written by drawing the free body diagrams for the lumped masses as shown in Fig. 1.4, Fig. 1.5, Fig. 1.6 and Fig. 1.7. The equilibrium equations can be written as follows:

The equation of equilibrium of the first floor in Building 1 that is connected with the first floor in Building 2 is as follows:

$$m_{11}\ddot{y}_{11} + c_{11}\left(\dot{y}_{11} - \dot{y}_{B1}\right) + k_{11}\left(y_{11} - y_{B1}\right) - c_{12}\left(\dot{y}_{12} - \dot{y}_{11}\right) - k_{12}\left(y_{12} - y_{11}\right)$$
$$+ c_{11,21}\left(\dot{y}_{11,B1} - \dot{y}_{21,B2}\right) + k_{11,21}\left(y_{11,B1} - y_{21,B2}\right) = F_{11}\left(t\right) \tag{1.1}$$

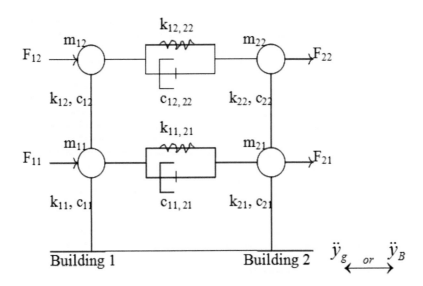

Fig. 1.3 Schematic diagram of the two adjacent 2-DOF fixed base systems.

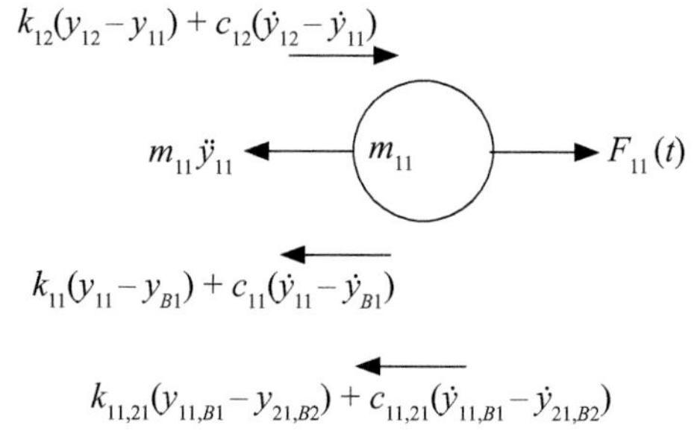

$$k_{12}(y_{12} - y_{11}) + c_{12}(\dot{y}_{12} - \dot{y}_{11})$$

$$m_{11}\ddot{y}_{11} \longleftarrow \left(m_{11} \right) \longrightarrow F_{11}(t)$$

$$k_{11}(y_{11} - y_{B1}) + c_{11}(\dot{y}_{11} - \dot{y}_{B1})$$

$$k_{11,21}(y_{11,B1} - y_{21,B2}) + c_{11,21}(\dot{y}_{11,B1} - \dot{y}_{21,B2})$$

Fig. 1.4 Free body diagram for lumped mass m_{11} of first floor of Building 1.

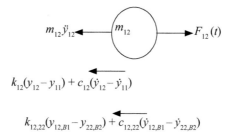

$$m_{12}\ddot{y}_{12} \longleftarrow \left(m_{12} \right) \longrightarrow F_{12}(t)$$

$$k_{12}(y_{12} - y_{11}) + c_{12}(\dot{y}_{12} - \dot{y}_{11})$$

$$k_{12,22}(y_{12,B1} - y_{22,B2}) + c_{12,22}(\dot{y}_{12,B1} - \dot{y}_{22,B2})$$

Fig. 1.5 Free body diagram for lumped mass m_{12} of second floor of Building 1.

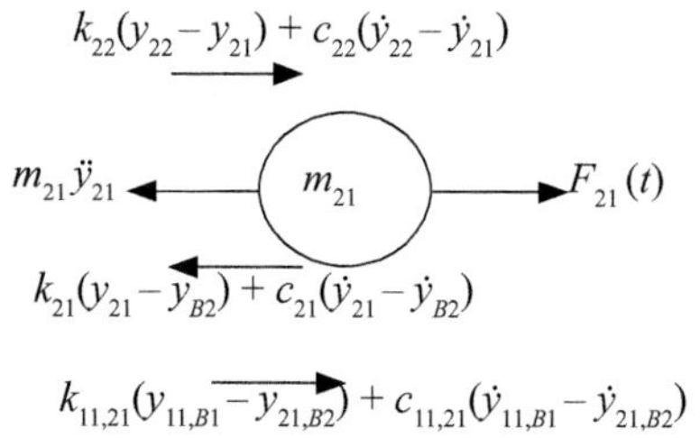

$$k_{22}(y_{22} - y_{21}) + c_{22}(\dot{y}_{22} - \dot{y}_{21})$$

$$m_{21}\ddot{y}_{21} \longleftarrow \left(m_{21} \right) \longrightarrow F_{21}(t)$$

$$k_{21}(y_{21} - y_{B2}) + c_{21}(\dot{y}_{21} - \dot{y}_{B2})$$

$$k_{11,21}(y_{11,B1} - y_{21,B2}) + c_{11,21}(\dot{y}_{11,B1} - \dot{y}_{21,B2})$$

Fig. 1.6 Free body diagram for lumped mass m_{21} of first floor of Building 2.

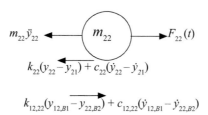

$$m_{22}\ddot{y}_{22} \longleftarrow \left(m_{22} \right) \longrightarrow F_{22}(t)$$

$$k_{22}(y_{22} - y_{21}) + c_{22}(\dot{y}_{22} - \dot{y}_{21})$$

$$k_{12,22}(y_{12,B1} - y_{22,B2}) + c_{12,22}(\dot{y}_{12,B1} - \dot{y}_{22,B2})$$

Fig. 1.7 Free body diagram for lumped mass m_{22} of second floor of Building 2.

Here, $\ddot{y}_{11} = \ddot{y}_{11,B1} + \ddot{y}_{B1}$ = absolute acceleration

$\dot{y}_{11} - \dot{y}_{B1} = \dot{y}_{11,B1}$ = relative velocity with respect to base

$y_{11} - y_{B1} = y_{11,B1}$ = relative displacement with respect to base

Substituting for the above relations in Eq. (1.1) and rearranging

$$m_{11}\ddot{y}_{11,B1} + c_{11}\left(\dot{y}_{11,B1}\right) + k_{11}\left(y_{11,B1}\right) - c_{12}\left(\dot{y}_{12,B1} - \dot{y}_{11,B1}\right) - k_{12}\left(y_{12,B1} - y_{11,B1}\right)$$
$$+c_{11,21}\left(\dot{y}_{11,B1} - \dot{y}_{21,B2}\right) + k_{11,21}\left(y_{11,B1} - y_{21,B2}\right) = F_{11}\left(t\right) - m_{11}\ddot{y}_{B1} \tag{1.2}$$

Similarly, for the second level in Building 1, the equation obtained is:

$$m_{12}\ddot{y}_{12} + c_{12}\left(\dot{y}_{12} - \dot{y}_{11}\right) + k_{12}\left(y_{12} - y_{11}\right) + c_{12,22}\left(\dot{y}_{12,B1} - \dot{y}_{22,B2}\right)$$
$$+k_{12,22}\left(y_{12,B1} - y_{22,B2}\right) = F_{12}\left(t\right) \tag{1.3}$$

By substituting for absolute acceleration and relative velocity and displacement before rearranging, the equation of motion can be rewritten as

$$m_{12}\ddot{y}_{12,B1} + c_{12}\left(\dot{y}_{12,B1} - \dot{y}_{11,B1}\right) + k_{12}\left(y_{12,B1} - y_{11,B1}\right) + c_{12,22}\left(\dot{y}_{12,B1} - \dot{y}_{22,B2}\right)$$
$$+k_{12,22}\left(y_{12,B1} - y_{22,B2}\right) = F_{12}\left(t\right) - m_{12}\ddot{y}_{B1} \tag{1.4}$$

The equation of motion for adjacent building models is obtained from the equilibrium equations of the free body diagram of each of the lumped mass of buildings. Thus, for each of the buildings, Eqs. (1.2) and (1.4) can be written in matrix form. The dynamic equation in matrix form for Building 1 is as follows:

$$\begin{bmatrix} m_{11} & 0 \\ 0 & m_{12} \end{bmatrix}\begin{bmatrix} \ddot{y}_{11,B1} \\ \ddot{y}_{12,B1} \end{bmatrix} + \begin{bmatrix} c_{11}+c_{12} & -c_{12} \\ -c_{12} & c_{12} \end{bmatrix}\begin{bmatrix} \dot{y}_{11,B1} \\ \dot{y}_{12,B1} \end{bmatrix} + \begin{bmatrix} k_{11}+k_{12} & -k_{12} \\ -k_{12} & k_{12} \end{bmatrix}\begin{bmatrix} y_{11,B1} \\ y_{12,B1} \end{bmatrix}$$
$$+\begin{bmatrix} c_{11,21}\left(\dot{y}_{11,B1} - \dot{y}_{21,B2}\right) + k_{11,21}\left(y_{11,B1} - y_{21,B2}\right) \\ c_{12,22}\left(\dot{y}_{12,B1} - \dot{y}_{22,B2}\right) + k_{12,22}\left(y_{12,B1} - y_{22,B2}\right) \end{bmatrix} = \begin{bmatrix} F_{11} \\ F_{12} \end{bmatrix} - \begin{bmatrix} m_{11} & 0 \\ 0 & m_{12} \end{bmatrix}\begin{bmatrix} \ddot{y}_{B1} \\ \ddot{y}_{B1} \end{bmatrix} \tag{1.5}$$

The equation of equilibrium of the first floor in Building 2 is similar to that in Building 1. In these equations, y_{11}, y_{12}, y_{21} and y_{22} denote the absolute displacements of the lumped mass in Buildings 1 and 2, respectively. Here, the first subscript denotes the building number while the second denotes the nth lumped floor of the building. The $y_{11,B1}, y_{12,B1}, y_{21,B2}$ and $y_{22,B2}$ in these equations are the relative displacements of floors with respect to the base of Buildings 1 and 2, respectively; $\ddot{y}_{B1}$ and $\ddot{y}_{B2}$ denote the acceleration of the base or the ground acceleration which each of the two buildings is subjected to. Since the spatial variation of the earthquake is not considered, these accelerations will be equal. A convenient matrix form can be developed by first combining these equations that lead to the expression:

$$
\begin{bmatrix} m_{11} & 0 & 0 & 0 \\ 0 & m_{12} & 0 & 0 \\ 0 & 0 & m_{21} & 0 \\ 0 & 0 & 0 & m_{22} \end{bmatrix} \begin{bmatrix} \ddot{y}_{11,B1} \\ \ddot{y}_{12,B1} \\ \ddot{y}_{21,B2} \\ \ddot{y}_{22,B2} \end{bmatrix} + \begin{bmatrix} c_{11}+c_{12} & -c_{12} & 0 & 0 \\ -c_{12} & c_{12} & 0 & 0 \\ 0 & 0 & c_{21}+c_{22} & -c_{22} \\ 0 & 0 & -c_{22} & c_{22} \end{bmatrix} \begin{bmatrix} \dot{y}_{11,B1} \\ \dot{y}_{12,B1} \\ \dot{y}_{21,B2} \\ \dot{y}_{22,B2} \end{bmatrix}
$$

$$
+ \begin{bmatrix} k_{11}+k_{12} & -k_{12} & 0 & 0 \\ -k_{12} & k_{12} & 0 & 0 \\ 0 & 0 & k_{21}+k_{22} & -k_{22} \\ 0 & 0 & -k_{22} & k_{22} \end{bmatrix} \begin{bmatrix} y_{11,B1} \\ y_{12,B1} \\ y_{21,B2} \\ y_{22,B2} \end{bmatrix}
$$

$$
+ \begin{bmatrix} c_{11,21}\left(\dot{y}_{11,B1}-\dot{y}_{21,B2}\right)+k_{11,21}\left(y_{11,B1}-y_{21,B2}\right) \\ c_{12,22}\left(\dot{y}_{12,B1}-\dot{y}_{22,B2}\right)+k_{12,22}\left(y_{12,B1}-y_{22,B2}\right) \\ -c_{11,21}\left(\dot{y}_{11,B1}-\dot{y}_{21,B2}\right)-k_{11,21}\left(y_{11,B1}-y_{21,B2}\right) \\ -c_{12,22}\left(\dot{y}_{12,B1}-\dot{y}_{22,B2}\right)-k_{12,22}\left(y_{12,B1}-y_{22,B2}\right) \end{bmatrix} \tag{1.6}
$$

$$
= \begin{bmatrix} F_{11} \\ F_{12} \\ F_{21} \\ F_{22} \end{bmatrix} - \begin{bmatrix} m_{11} & 0 & 0 & 0 \\ 0 & m_{12} & 0 & 0 \\ 0 & 0 & m_{21} & 0 \\ 0 & 0 & 0 & m_{22} \end{bmatrix} \begin{bmatrix} \ddot{y}_{B1} \\ \ddot{y}_{B1} \\ \ddot{y}_{B2} \\ \ddot{y}_{B2} \end{bmatrix}
$$

Finally, for collecting stiffness and damping contributions, the equation of motion for adjacent building systems as illustrated in Fig. 1.3 can be rewritten as:

$$
\begin{bmatrix} m_{11} & 0 & 0 & 0 \\ 0 & m_{12} & 0 & 0 \\ 0 & 0 & m_{21} & 0 \\ 0 & 0 & 0 & m_{22} \end{bmatrix} \begin{bmatrix} \ddot{y}_{11,B1} \\ \ddot{y}_{12,B1} \\ \ddot{y}_{21,B2} \\ \ddot{y}_{22,B2} \end{bmatrix}
$$

$$
+ \left\{ \begin{bmatrix} c_{11}+c_{12} & -c_{12} & 0 & 0 \\ -c_{12} & c_{12} & 0 & 0 \\ 0 & 0 & c_{21}+c_{22} & -c_{22} \\ 0 & 0 & -c_{22} & c_{22} \end{bmatrix} + \begin{bmatrix} c_{11,21} & 0 & -c_{11,21} & 0 \\ 0 & c_{12,22} & 0 & -c_{12,22} \\ -c_{11,21} & 0 & c_{11,21} & 0 \\ 0 & -c_{12,22} & 0 & c_{12,22} \end{bmatrix} \right\} \begin{bmatrix} \dot{y}_{11,B1} \\ \dot{y}_{12,B1} \\ \dot{y}_{21,B2} \\ \dot{y}_{22,B2} \end{bmatrix}
$$

$$
+ \left\{ \begin{bmatrix} k_{11}+k_{12} & -k_{12} & 0 & 0 \\ -k_{12} & k_{12} & 0 & 0 \\ 0 & 0 & k_{21}+k_{22} & -k_{22} \\ 0 & 0 & -k_{22} & k_{22} \end{bmatrix} + \begin{bmatrix} k_{11,21} & 0 & -k_{11,21} & 0 \\ 0 & k_{12,22} & 0 & -k_{12,22} \\ -k_{11,21} & 0 & k_{11,21} & 0 \\ 0 & -k_{12,22} & 0 & k_{12,22} \end{bmatrix} \right\} \begin{bmatrix} y_{11,B1} \\ y_{12,B1} \\ y_{21,B2} \\ y_{22,B2} \end{bmatrix} \tag{1.7}
$$

$$
= \begin{bmatrix} F_{11} \\ F_{12} \\ F_{21} \\ F_{22} \end{bmatrix} - \begin{bmatrix} m_{11} & 0 & 0 & 0 \\ 0 & m_{12} & 0 & 0 \\ 0 & 0 & m_{21} & 0 \\ 0 & 0 & 0 & m_{22} \end{bmatrix} \begin{bmatrix} \ddot{y}_{B1} \\ \ddot{y}_{B1} \\ \ddot{y}_{B2} \\ \ddot{y}_{B2} \end{bmatrix}
$$

In a two-degree of freedom model, the equation of motion for adjacent building systems at both the floor levels is seen in Eq. (1.7). As in the case of two-degree of freedom system, the equation of motion for adjacent buildings which are modelled as multi-degree of freedom systems can be obtained by writing the equilibrium equations from the free body diagram of each of the lumped mass of the building.

1.7 Equations of Motion

Building A and Building B have $n + m$ storeys and n storeys, respectively, as shown in Fig. 1.8. The mass, damping coefficient and shear stiffness values for the ith storey are $m_{i,1}$, $c_{i,1}$, $k_{i,1}$ for Building A and $m_{i,2}$, $c_{i,2}$, $k_{i,2}$ for Building B, respectively. The stiffness of viscous damper and the coefficient of damping on the first floor are represented as $k_{d,i}$ and $c_{d,i}$, respectively.

The dynamic model of coupled buildings is taken to have a $2n+m$ degree of freedom system. The equations of motion for this system are expressed as

$$M\ddot{Y} + (C + C_d)\dot{Y} + (K + K_d)Y = -MI\ddot{Y}_g \tag{1.8}$$

where M, C and K are the mass, damping and stiffness matrices of the coupled buildings, respectively; C_d and K_d are the additional damping and stiffness matrices consisting the installation of the fluid viscous damper; Y is the relative displacement vector with respect to the ground and consists of Building A's displacements in the first $n + m$ positions and Building B's displacements in the last n positions; I is a unity matrix with all its diagonal elements equal to unity and the rest equal to 0; $\ddot{y}_g$ is the earthquake acceleration at the foundations of the buildings. The details of each matrix are shown as follows:

$$M = \begin{bmatrix} m_{n+m,n+m} & 0_{n+m,n} \\ 0_{n,n+m} & m_{n,n} \end{bmatrix}; K = \begin{bmatrix} K_{n+m,n+m} & 0_{n+m,n} \\ 0_{n,n+m} & K_{n,n} \end{bmatrix}; C = \begin{bmatrix} C_{n+m,n+m} & 0_{n+m,n} \\ 0_{n,n+m} & C_{n,n} \end{bmatrix} \tag{1.9}$$

$$C_d = \begin{bmatrix} c_{d(n,n)} & 0_{(n,m)} & -c_{d(n,n)} \\ 0_{(m,n)} & 0_{(m,m)} & 0_{(m,n)} \\ -c_{d(n,n)} & 0_{(n,m)} & c_{d(n,n)} \end{bmatrix}; K_d = \begin{bmatrix} k_{d(n,n)} & 0_{(n,m)} & -k_{d(n,n)} \\ 0_{(m,n)} & 0_{(m,m)} & 0_{(m,n)} \\ -k_{d(n,n)} & 0_{(n,m)} & k_{d(n,n)} \end{bmatrix} \tag{1.10}$$

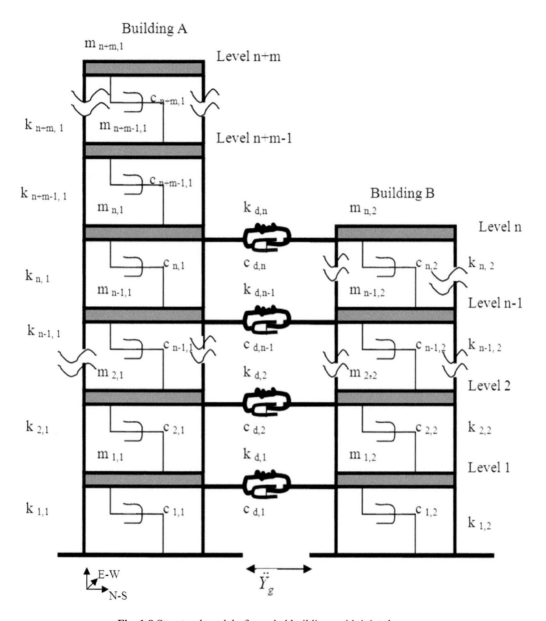

Fig. 1.8 Structural model of coupled buildings with joint dampers.

$$m_{n+m,n+m} = \begin{bmatrix} m_{11} & & & & \\ & m_{21} & & & \\ & & \cdot & & \\ & & & \cdot & \\ & & & m_{n+m-1,1} & \\ & & & & m_{n+m,1} \end{bmatrix} ; m_{n,n} = \begin{bmatrix} m_{12} & & & & \\ & m_{22} & & & \\ & & \cdot & & \\ & & & \cdot & \\ & & & m_{n-1,2} & \\ & & & & m_{n,2} \end{bmatrix}$$ (1.11)

$$k_{n+m,n+m} = \begin{bmatrix} k_{11}+k_{21} & -k_{21} & & & \\ -k_{21} & k_{21}+k_{31} & -k_{31} & & \\ & & \cdot & & \\ & & & \cdot & \\ & & -k_{n+m-1,1} & k_{n+m-1,1}+k_{n+m,1} & -k_{n+m,1} \\ & & & -k_{n+m,1} & k_{n+m,1} \end{bmatrix}$$ (1.12)

$$k_{n,n} = \begin{bmatrix} k_{12}+k_{22} & -k_{22} & & & \\ -k_{22} & k_{22}+k_{32} & -k_{32} & & \\ & & \cdot & & \\ & & & \cdot & \\ & & -k_{n-1,2} & k_{n-1,2}+k_{n,2} & -k_{n,2} \\ & & & -k_{n,2} & k_{n,2} \end{bmatrix}$$ (1.13)

$$c_{n+m,n+m} = \begin{bmatrix} c_{11}+c_{21} & -c_{21} & & & \\ -c_{21} & c_{21}+c_{31} & -c_{31} & & \\ & & \cdot & & \\ & & & \cdot & \\ & & -c_{n+m-1,1} & c_{n+m-1,1}+c_{n+m,1} & -c_{n+m,1} \\ & & & -c_{n+m,1} & c_{n+m,1} \end{bmatrix}$$ (1.14)

$$c_{n,n} = \begin{bmatrix} c_{12}+c_{22} & -c_{22} & & & \\ -c_{22} & c_{22}+c_{32} & -c_{32} & & \\ & & \cdot & & \\ & & & \cdot & \\ & & -c_{n-1,2} & c_{n-1,2}+c_{n,2} & -c_{n,2} \\ & & & -c_{n,2} & c_{n,2} \end{bmatrix}$$ (1.15)

$$c_{d(n,n)} = \begin{bmatrix} c_{d1} & & & & & \\ & c_{d2} & & & & \\ & & \cdot & & & \\ & & & \cdot & & \\ & & & & c_{d,n-1} & \\ & & & & & c_{dn} \end{bmatrix} ; k_{d(n,n)} = \begin{bmatrix} k_{d1} & & & & & \\ & k_{d2} & & & & \\ & & \cdot & & & \\ & & & \cdot & & \\ & & & & k_{d,n-1} & \\ & & & & & k_{dn} \end{bmatrix} \qquad (1.16)$$

$$Y^T = \begin{bmatrix} y_{11}, y_{21}, \ldots, y_{n+m-1,1}, y_{n+m,1}, y_{12}, y_{22}, \ldots, y_{n-1,2}, y_{n,2} \end{bmatrix} \qquad (1.17)$$

And 0 in Eqs. (1.9) and (1.10) is described as a zero matrix. For time-domain analysis, the above equations can be used directly for any given time history record of ground motion. For the frequency domain analysis, the traditional random vibration-based SRSS (the square root of the sum of the squares of modal responses) method is used because of the classical damping properties of the damper-building system.

1.8 Outline of the Book

This book investigates the pounding effects for base-isolated buildings, the soil structure interaction effects on adjacent buildings considering impact effect and the seismic protection of adjacent buildings and the mitigation of earthquake-induced vibrations of the two adjacent structures. The passive control strategy of coupled buildings is investigated for seismic protection with comparison to active and semi-active control strategies. To achieve these objectives, this book is divided into twelve chapters (including the introduction), which are organised as follows:

Chapter 1 presents the potential of pounding reduction devices, providing a comprehensive review of topics that form the background of this book. In order to understand the effect of dampers on the dynamic characteristics of building models, with respect to shear wave velocity on SSI systems, the available modal analysis methods for the SSI systems and its application to MDOF modal response history analyses are critically reviewed. The main differences and advantages are pointed out. Optimal design of passive, active and semi-active devices in structural control optimisation and design approaches are also reviewed.

Chapter 2 contains a formulation of multi-degree-of-freedom adjacent buildings used for analytical study of the book. Formulations for the base-isolated buildings and fixed buildings with SSI systems under the effect of impact are introduced in equations for adjacent buildings modelled under inelastic and elastic systems. The reference buildings without the SSI effects are described in order to investigate the effect of SSI on the behaviour pounding of the system.

Chapter 3 shows the principles and dynamics of controlled passive-active and semi-active control concepts introduced along with the state-space concept, feedback control systems and stability of the active and semi-active control structures.

Chapter 4 provides the control algorithms to design control systems. The basic components of some control algorithms, such as H_2, H_∞, H_2/LQG, LQR with output feedback, clipped optimal control and Fuzzy Logic Control (FLC) are described here.

Chapter 5 contains the optimum design using genetic algorithm, NSGA-II and Pareto optimal solution, having discussed control algorithms. Basic mechanisms, components and advantages of genetic algorithms (GAs) and multi-objective genetic algorithm as a structural optimisation method are defined in this section.

Chapter 6 shows a numerical study of two adjacent buildings on damping control, solving two different methods for the equation of motion, such as the rigorous method using direct integration method and the proposed approximation method, using MDOF modal response history analysis. The response histories of the eight degrees of freedom for each vibration mode shapes of the adjacent buildings modelled in various SSI effects are described, including the applications and enhancements of particular adjacent building models.

The results for response analysis and time-history analysis are shown in Chapter 7. Additionally, the benefits of installing passive, active and semi-active control systems are evaluated for adjacent buildings. This chapter summarises the findings of this book, which includes the coupled buildings connected by passive dampers evaluated through the values of changing shear wave velocity on the behaviour of coupled buildings under various SSI systems in two-directional seismic ground motions. Additionally, several cases are defined in Chapter 7. The results of a number of these cases are presented to demonstrate the approach and the potential effectiveness of this methodology.

Chapter 8 proposes discussions and conclusions. A list of References and Appendices sum up the chapter.

2

Mathematical Modelling of Adjacent Buildings under Earthquake Loading

2.1 Introduction

Successful studies have been undertaken to investigate the pounding effect on buildings and to develop engineering solutions regarding the behaviour of buildings under different soil effects. However, since the last two decades, the pounding effect of adjacent buildings considering the soil-structure effects is yet to be understood fully. This chapter has two parts describing the proposed system models—firstly, the behaviour of the symmetric adjacent buildings having base isolations is investigated with pounding effects, ignoring the soil structure interaction (SSI) system and torsion effect; next, in order to consider the effects of the large and small SSI systems, two-way asymmetric coupled buildings are modelled in Section 2.5. The objective of the first part is to validate the equation of motion for the SSI system compared to the multi-degrees of freedom modal equations of motion.

2.2 Calculation of Building Separation Distance

The required distance between two adjacent buildings in order to reduce the risk of seismic pounding was determined by Valles and Reinhorn (1997), as shown Fig. 2.1. For adjacent buildings, the seismic pounding problems, which have received increasing attention, were considered by many researchers. The International Building Code (2003) calculated the required distance between two adjacent buildings in terms of the square root of the sum

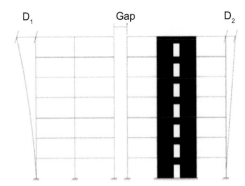

Fig. 2.1 Definition of gap International Building Code (2003).

of squares (SRSS) of the individual building displacements due to the risk of seismic pounding. D1 and D2 in Fig. 2.1 are displacements of the adjacent buildings as shown at δ_{M1} and δ_{M2}. The following standard is the concept of specification from International Building Code (2003).

The aforementioned research studies focus on avoidance of mutual pounding between adjacent buildings, which are spaced very closely. In this book, the alleviation of earthquake responses of coupled buildings, spaced with a definite distance by using fluid dampers to connect them, is presented. Previous studies demonstrate the effectiveness of passive energy-dispersing systems to develop seismic performance of connected buildings observed through extensive analytical and experimental investigations. However, the performance of fluid viscous dampers in terms of the reduction of displacement, acceleration and shear force responses of adjacent buildings, determining the optimal design parameters of dampers for adjacent buildings of the same stiffness ratios and different heights have not been investigated fully. A formulation of multi-degree of freedom equations of motion for viscous dampers connecting adjacent high-rise buildings under earthquake excitations is separately presented. A time history analysis and response spectrum analysis are performed for two adjacent buildings to find out the dynamic response of the structures.

2.3 Torsional Response of Asymmetric System

The torsion irregularity existing in multi-storey or buildings is generally caused due to hinged links used to connect two neighbouring floors or non-symmetric form of the plan geometry or the load-carrying member's rigidity distribution. In previous studies, the lumped mass models were categorised into models that either include or neglect the consideration of building torsion. Some researchers have ignored the rotation of the building even though torsion occurs due to damper devices used for interconnecting two buildings. The rotation of buildings is generally quite small when compared to the lateral components. Westermo (1989) investigated the dynamic implications of connecting closely adjacent structures by a hinged beam system for the purpose of reducing the risk of pounding under earthquake excitations and also suggested use of hinged links to connect two neighbouring floors, if the neighbouring floors were in alignment. It is understandable that use of damping devices between two adjacent buildings can reduce the chances of pounding, but it changes the dynamic characteristics of the separated buildings, augments undesirable torsion action due to non-symmetric form of the plan geometry or the load-carrying member's rigidity distribution, and increases the base shear of the stiffer building.

Uz and Hadi (2009) showed that fluid viscous dampers located in one direction are not very effective in reducing the earthquake responses of the adjacent buildings when selected earthquakes are applied in two directions. Thereby, in this study of Uz and Hadi (2009), in order to carry out the use of dampers for two directions under strong earthquakes, the analysis is investigated in both directions in the structural responses of two neighbouring buildings, which have the same stiffness ratios and different heights, connected with two different damper parameters under various earthquake excitations in each model. Although two buildings are assumed as symmetric buildings in their plane of alignment, torsion effects can occur because of linked viscous dampers between them. However, symmetric adjacent buildings with torsion effects are considered in Section 2.5.

2.4 Equation of Motion for the Effect of Pounding

As shown in Fig. 2.2, the adjacent buildings have been modelled as four- and three-storeyed buildings. In order to investigate the behaviour of colliding base-isolated buildings, a three-dimensional model with the help of each storey's mass lumped on the floor level was conducted. The dynamic equation of motion for the two base-isolated buildings is expressed in Eq. 2.1, including the pounding involved responses of base-isolated buildings modelled with inelastic systems at each floor level as:

$$
\begin{bmatrix} M_1 & 0 & 0 \\ 0 & M_1 & 0 \\ 0 & 0 & M_1 \end{bmatrix} \begin{bmatrix} \ddot{X} \\ \ddot{Y} \\ \ddot{Z} \end{bmatrix} + \begin{bmatrix} C_x & 0 & 0 \\ 0 & C_y & 0 \\ 0 & 0 & C_z \end{bmatrix} \begin{bmatrix} \dot{X} \\ \dot{Y} \\ \dot{Z} \end{bmatrix} + \begin{bmatrix} F_x^S \\ F_y^S \\ F_z^S \end{bmatrix} + \begin{bmatrix} F_x^P \\ F_y^P \\ F_z^P \end{bmatrix}
$$
$$
= \begin{bmatrix} F_x \\ F_y \\ F_z \end{bmatrix} - \begin{bmatrix} M_2 & 0 & 0 \\ 0 & M_2 & 0 \\ 0 & 0 & M_2 \end{bmatrix} \begin{bmatrix} \ddot{X} \\ \ddot{Y} \\ \ddot{Z} \end{bmatrix} - \begin{bmatrix} M_3 & 0 & 0 \\ 0 & M_3 & 0 \\ 0 & 0 & M_3 \end{bmatrix} \begin{bmatrix} \ddot{X}_g \\ \ddot{Y}_g \\ \ddot{Z}_g \end{bmatrix} \tag{2.1}
$$

It is assumed that the two buildings may not remain in the linear elastic range and the buildings may reach the storey yield strength to be inelastic under the considered earthquake

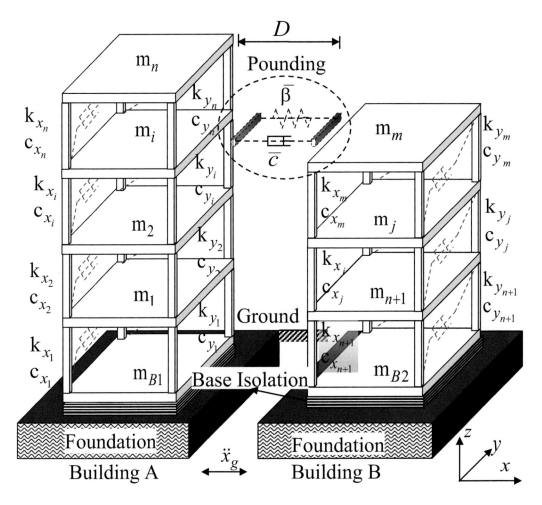

Fig. 2.2 Three-dimensional model of colliding base-isolated adjacent buildings.

excitation (Catal 2002). Hence, an elastic-plastic approximation of the storey drift-shear force relation has been fulfilled for the longitudinal (x) and transverse (y) directions, whereas the two buildings are assumed to be in the linear elastic range for the vertical direction (z). For i = 1, 2, 3, 4, 5, 6, 7, masses, damping and stiffness coefficients of Building A and Building B in Fig. 2.2 in the longitudinal (x) and transverse (y) directions are shown as m_{xi}, c_{xi}, k_{xi}, m_{yi}, c_{yi}, and k_{yi}, respectively. The m_{Bi} (i = 1, 2) denotes the mass of the base of both the buildings, respectively.

$$
\ddot{X} = \begin{bmatrix} \ddot{x}_1 \\ \ddot{x}_2 \\ \ddot{x}_3 \\ \ddot{x}_4 \\ \ddot{x}_5 \\ \ddot{x}_6 \\ \ddot{x}_7 \\ \ddot{x}_{B1} \\ \ddot{x}_{B2} \end{bmatrix}
\ddot{Y} = \begin{bmatrix} \ddot{y}_1 \\ \ddot{y}_2 \\ \ddot{y}_3 \\ \ddot{y}_4 \\ \ddot{y}_5 \\ \ddot{y}_6 \\ \ddot{y}_7 \\ \ddot{y}_{B1} \\ \ddot{y}_{B2} \end{bmatrix}
\ddot{Z} = \begin{bmatrix} \ddot{z}_1 \\ \ddot{z}_2 \\ \ddot{z}_3 \\ \ddot{z}_4 \\ \ddot{z}_5 \\ \ddot{z}_6 \\ \ddot{z}_7 \\ \ddot{z}_{B1} \\ \ddot{z}_{B2} \end{bmatrix}
\dot{X} = \begin{bmatrix} \dot{x}_1 \\ \dot{x}_2 \\ \dot{x}_3 \\ \dot{x}_4 \\ \dot{x}_5 \\ \dot{x}_6 \\ \dot{x}_7 \\ \dot{x}_{B1} \\ \dot{x}_{B2} \end{bmatrix}
\dot{Y} = \begin{bmatrix} \dot{y}_1 \\ \dot{y}_2 \\ \dot{y}_3 \\ \dot{y}_4 \\ \dot{y}_5 \\ \dot{y}_6 \\ \dot{y}_7 \\ \dot{y}_{B1} \\ \dot{y}_{B2} \end{bmatrix}
\dot{Z} = \begin{bmatrix} \dot{z}_1 \\ \dot{z}_2 \\ \dot{z}_3 \\ \dot{z}_4 \\ \dot{z}_5 \\ \dot{z}_6 \\ \dot{z}_7 \\ \dot{z}_{B1} \\ \dot{z}_{B2} \end{bmatrix}
\tag{2.2a–f}
$$

$$
X = \begin{bmatrix} x_1 \\ x_2 \\ x_3 \\ x_4 \\ x_5 \\ x_6 \\ x_7 \\ x_{B1} \\ x_{B2} \end{bmatrix}
Y = \begin{bmatrix} y_1 \\ y_2 \\ y_3 \\ y_4 \\ y_5 \\ y_6 \\ y_7 \\ y_{B1} \\ y_{B2} \end{bmatrix}
Z = \begin{bmatrix} z_1 \\ z_2 \\ z_3 \\ z_4 \\ z_5 \\ z_6 \\ z_7 \\ z_{B1} \\ z_{B2} \end{bmatrix}
\ddot{X}_g = \begin{bmatrix} \ddot{x}_g \\ \ddot{x}_g \\ \ddot{x}_g \\ \ddot{x}_g \\ \ddot{x}_g \\ \ddot{x}_g \\ \ddot{x}_g \\ 0 \\ 0 \end{bmatrix}
\ddot{Y}_g = \begin{bmatrix} \ddot{y}_g \\ \ddot{y}_g \\ \ddot{y}_g \\ \ddot{y}_g \\ \ddot{y}_g \\ \ddot{y}_g \\ \ddot{y}_g \\ 0 \\ 0 \end{bmatrix}
\ddot{Z}_g = \begin{bmatrix} \ddot{z}_g \\ \ddot{z}_g \\ \ddot{z}_g \\ \ddot{z}_g \\ \ddot{z}_g \\ \ddot{z}_g \\ \ddot{z}_g \\ 0 \\ 0 \end{bmatrix}
\tag{2.3a–f}
$$

Here $\ddot{x}_i$, $\dot{x}_i$, x_i, $\ddot{y}_i$, $\dot{y}_i$, y_i, $\ddot{z}_i$, $\dot{z}_i$ and z_i ($i = 1, 2, 3, 4, 5, 6, 7$) are the relative acceleration, velocity and displacement of a single storey with respect to the base of the structure in the longitudinal, transverse and vertical directions, respectively. The $\ddot{x}_g$, $\ddot{y}_g$ and $\ddot{z}_g$ in Eqs. 2.3a–f are earthquake accelerations in the longitudinal, transverse and vertical directions. The El-Centro (1940) and Turkey (1999) earthquake records are used to examine the seismic response of two buildings under different ground motions in this part of book.

$$
M_1 = \begin{bmatrix}
m_1 & 0 & 0 & 0 & 0 & 0 & 0 & 0 & 0 \\
0 & m_2 & 0 & 0 & 0 & 0 & 0 & 0 & 0 \\
0 & 0 & m_3 & 0 & 0 & 0 & 0 & 0 & 0 \\
0 & 0 & 0 & m_4 & 0 & 0 & 0 & 0 & 0 \\
0 & 0 & 0 & 0 & m_5 & 0 & 0 & 0 & 0 \\
0 & 0 & 0 & 0 & 0 & m_6 & 0 & 0 & 0 \\
0 & 0 & 0 & 0 & 0 & 0 & m_7 & 0 & 0 \\
a_1{}^* m_1 & a_1{}^* m_2 & a_1{}^* m_3 & a_1{}^* m_4 & 0 & 0 & 0 & m_1+m_2+m_3+m_4+m_{B1} & 0 \\
0 & 0 & 0 & 0 & a_2{}^* m_5 & a_2{}^* m_6 & a_2{}^* m_7 & 0 & m_5+m_6+m_7+m_{B2}
\end{bmatrix}
\tag{2.4}
$$

$$
M_2 = \begin{bmatrix}
0 & 0 & 0 & 0 & 0 & 0 & 0 & m_1 & 0 \\
0 & 0 & 0 & 0 & 0 & 0 & 0 & m_2 & 0 \\
0 & 0 & 0 & 0 & 0 & 0 & 0 & m_3 & 0 \\
0 & 0 & 0 & 0 & 0 & 0 & 0 & m_4 & 0 \\
0 & 0 & 0 & 0 & 0 & 0 & 0 & 0 & m_5 \\
0 & 0 & 0 & 0 & 0 & 0 & 0 & 0 & m_6 \\
0 & 0 & 0 & 0 & 0 & 0 & 0 & 0 & m_7 \\
0 & 0 & 0 & 0 & 0 & 0 & 0 & 0 & 0 \\
0 & 0 & 0 & 0 & 0 & 0 & 0 & 0 & 0
\end{bmatrix}, M_3 = \begin{bmatrix}
m_1 & 0 & 0 & 0 & 0 & 0 & 0 & 0 & 0 \\
0 & m_2 & 0 & 0 & 0 & 0 & 0 & 0 & 0 \\
0 & 0 & m_3 & 0 & 0 & 0 & 0 & 0 & 0 \\
0 & 0 & 0 & m_4 & 0 & 0 & 0 & 0 & 0 \\
0 & 0 & 0 & 0 & m_5 & 0 & 0 & 0 & 0 \\
0 & 0 & 0 & 0 & 0 & m_6 & 0 & 0 & 0 \\
0 & 0 & 0 & 0 & 0 & 0 & m_7 & 0 & 0 \\
0 & 0 & 0 & 0 & 0 & 0 & 0 & 0 & 0 \\
0 & 0 & 0 & 0 & 0 & 0 & 0 & 0 & 0
\end{bmatrix} \quad (2.5a–b)
$$

Here m_i (i = 1, ..., 7) denotes mass of a single storey of both buildings, whereas mass of the base of two buildings is shown as m_{Bi} (i = 1, 2). Then a_1 and a_2 (a_i (i = 1, 2)) are defined to account for sliding in Building *A* and Building *B* in the longitudinal direction, respectively. Its value equals 0 for no sliding and 1 for sliding. When a_i equals zero, it implies zero sliding velocity and acceleration for the related building.

$$
C_x = \begin{bmatrix}
c_{x1}+c_{x2} & -c_{x2} & 0 & 0 & 0 & 0 & 0 & 0 & 0 \\
-c_{x2} & c_{x2}+c_{x3} & -c_{x3} & 0 & 0 & 0 & 0 & 0 & 0 \\
0 & -c_{x3} & c_{x3}+c_{x4} & -c_{x4} & 0 & 0 & 0 & 0 & 0 \\
0 & 0 & -c_{x4} & c_{x4} & 0 & 0 & 0 & 0 & 0 \\
0 & 0 & 0 & 0 & c_{x5}+c_{x6} & -c_{x6} & 0 & 0 & 0 \\
0 & 0 & 0 & 0 & -c_{x6} & c_{x6}+c_{x7} & -c_{x7} & 0 & 0 \\
0 & 0 & 0 & 0 & 0 & -c_{x7} & c_{x7} & 0 & 0 \\
0 & 0 & 0 & 0 & 0 & 0 & 0 & 0 & 0 \\
0 & 0 & 0 & 0 & 0 & 0 & 0 & 0 & 0
\end{bmatrix}
$$

$$
C_y = \begin{bmatrix}
c_{y1}+c_{y2} & -c_{y2} & 0 & 0 & 0 & 0 & 0 & 0 & 0 \\
-c_{y2} & c_{y2}+c_{y3} & -c_{y3} & 0 & 0 & 0 & 0 & 0 & 0 \\
0 & -c_{y3} & c_{y3}+c_{y4} & -c_{y4} & 0 & 0 & 0 & 0 & 0 \\
0 & 0 & -c_{y4} & c_{y4} & 0 & 0 & 0 & 0 & 0 \\
0 & 0 & 0 & 0 & c_{y5}+c_{y6} & -c_{y6} & 0 & 0 & 0 \\
0 & 0 & 0 & 0 & -c_{y6} & c_{y6}+c_{y7} & -c_{y7} & 0 & 0 \\
0 & 0 & 0 & 0 & 0 & -c_{y7} & c_{y7} & 0 & 0 \\
0 & 0 & 0 & 0 & 0 & 0 & 0 & 0 & 0 \\
0 & 0 & 0 & 0 & 0 & 0 & 0 & 0 & 0
\end{bmatrix} \quad (2.6a–c)
$$

$$
C_z = \begin{bmatrix}
c_{z1} + c_{z2} & -c_{z2} & 0 & 0 & 0 & 0 & 0 & 0 & 0 \\
-c_{z2} & c_{z2} + c_{z3} & -c_{z3} & 0 & 0 & 0 & 0 & 0 & 0 \\
0 & -c_{z3} & c_{z3} + c_{z4} & -c_{z4} & 0 & 0 & 0 & 0 & 0 \\
0 & 0 & -c_{z4} & c_{z4} & 0 & 0 & 0 & 0 & 0 \\
0 & 0 & 0 & 0 & c_{z5} + c_{z6} & -c_{z6} & 0 & 0 & 0 \\
0 & 0 & 0 & 0 & -c_{z6} & c_{z6} + c_{z7} & -c_{z7} & 0 & 0 \\
0 & 0 & 0 & 0 & 0 & -c_{z7} & c_{z7} & 0 & 0 \\
0 & 0 & 0 & 0 & 0 & 0 & 0 & 0 & 0 \\
0 & 0 & 0 & 0 & 0 & 0 & 0 & 0 & 0
\end{bmatrix}
$$

Here c_{xi}, c_{yi} and c_{zi} (i = 1,…,7) are the elastic structural damping coefficients in the longitudinal, transverse and vertical directions, respectively.

$$
F_x^S = \begin{bmatrix}
F_{x1}^S - F_{x2}^S \\
F_{x2}^S - F_{x3}^S \\
F_{x3}^S - F_{x4}^S \\
F_{x4}^S \\
F_{x5}^S - F_{x6}^S \\
F_{x6}^S - F_{x7}^S \\
F_{x7}^S \\
0 \\
0
\end{bmatrix}
\quad
F_y^S = \begin{bmatrix}
F_{y1}^S - F_{y2}^S \\
F_{y2}^S - F_{y3}^S \\
F_{y3}^S - F_{y4}^S \\
F_{y4}^S \\
F_{y5}^S - F_{y6}^S \\
F_{y6}^S - F_{y7}^S \\
F_{y7}^S \\
0 \\
0
\end{bmatrix}
\qquad (2.7a\text{–}c)
$$

$$
F_z^s = \begin{bmatrix}
k_{z1} + k_{z2} & -k_{z2} & 0 & 0 & 0 & 0 & 0 & 0 & 0 \\
-k_{z2} & k_{z2} + k_{z3} & -k_{z3} & 0 & 0 & 0 & 0 & 0 & 0 \\
0 & -k_{z3} & k_{z3} + k_{z4} & -k_{z4} & 0 & 0 & 0 & 0 & 0 \\
0 & 0 & -k_{z4} & k_{z4} & 0 & 0 & 0 & 0 & 0 \\
0 & 0 & 0 & 0 & k_{z5} + k_{z6} & -k_{z6} & 0 & 0 & 0 \\
0 & 0 & 0 & 0 & -k_{z6} & k_{z6} + k_{z7} & -k_{z7} & 0 & 0 \\
0 & 0 & 0 & 0 & 0 & -k_{z7} & k_{z7} & 0 & 0 \\
0 & 0 & 0 & 0 & 0 & 0 & 0 & 0 & 0 \\
0 & 0 & 0 & 0 & 0 & 0 & 0 & 0 & 0
\end{bmatrix}
\begin{bmatrix}
z_1 \\ z_2 \\ z_3 \\ z_4 \\ z_5 \\ z_6 \\ z_7 \\ z_{B1} \\ z_{B2}
\end{bmatrix}
$$

F_{xi}^s and F_{yi}^s (i = 1,…7) in Eqs. 2.7a–c are the inelastic storey shear forces for both elastic range and plastic range. If the storey shear forces are in the elastic range, F_{xi}^s and F_{yi}^s are equal to $k_{xi}(x_i - x_{i-1})$ and $k_{yi}(y_i - y_{i-1})$, respectively. When the storey yield strengths F_{xi}^y, F_{yi}^y are reached, F_{xi}^s and F_{yi}^s are equal to $\mp F_{xi}^y(t)$ and $\mp F_{yi}^y(t)$ for the plastic range in the longitudinal and transverse directions, respectively. In Eqs. 2.8a–c and 2.9a–c, the pounding force in the longitudinal direction, $F_{xij}^p(t)$ (i = 1, 2, 3, 4; j = 5, 6, 7), is arranged with the help of non-linear visco-elastic model according to the formula (Jankowski et al. 1998, Jankowski 2006, Jankowski 2008) as follows:

$$F_x^p(t) = \begin{bmatrix} F_{x15}^p(t) \\ F_{x26}^p(t) \\ F_{x37}^p(t) \\ 0 \\ -F_{x15}^p(t) \\ -F_{x26}^p(t) \\ -F_{x37}^p(t) \\ 0 \\ 0 \end{bmatrix}, \; F_y^p(t) = \begin{bmatrix} F_{y15}^p(t) \\ F_{y26}^p(t) \\ F_{y37}^p(t) \\ 0 \\ -F_{y15}^p(t) \\ -F_{y26}^p(t) \\ -F_{y37}^p(t) \\ 0 \\ 0 \end{bmatrix}, \; F_z^p(t) = \begin{bmatrix} F_{z15}^p(t) \\ F_{z26}^p(t) \\ F_{z37}^p(t) \\ 0 \\ -F_{z15}^p(t) \\ -F_{z26}^p(t) \\ -F_{z37}^p(t) \\ 0 \\ 0 \end{bmatrix} \quad \text{(2.8a–c)}$$

$$F_{xij}^p(t) = 0 \text{ for } \delta_{ij}(t) \le 0$$
$$F_{xij}^p(t) = \bar{\beta}\left(\delta_{ij}(t)\right)^{3/2} + \bar{c}_{ij}(t)\dot{\delta}_{ij}(t) \text{ for } \delta_{ij}(t) > 0 \text{ and } \dot{\delta}_{ij}(t) > 0 \quad \text{(2.9a–c)}$$
$$F_{xij}^p(t) = \bar{\beta}\left(\delta_{ij}(t)\right)^{3/2} \text{ for } \delta_{ij}(t) > 0 \text{ and } \dot{\delta}_{ij}(t) \le 0$$

$$\delta_{ij}(t) = u_i(t) - u_j(t) - D, \; \dot{\delta}_{ij}(t) = \dot{u}_i(t) - \dot{u}_j(t), \; \bar{c}_{ij}(t) = 2\bar{\xi}\sqrt{\bar{\beta}\sqrt{\delta_{ij}(t)}\frac{m_i m_j}{m_i + m_j}},$$

$$\bar{\xi} = \frac{9\sqrt{5}}{2}\frac{1-e^2}{e\left(e(9\pi-16)+16\right)} \quad \text{(2.10a–d)}$$

Here $\delta_{ij}(t)$ is the total relative displacement between the two buildings with respect to the foundation, $u_i(t)$ and $u_j(t)$ are the relative displacement of the *i*th floor of Building *A* and *j*th floor of Building *B* with respect to the foundation, respectively (Hadi and Uz 2010a, Uz and Hadi 2010). *D* is the initial gap between the buildings exposed to different ground motion excitations, which is verified to investigate the effect of different gap distances in the book; $\dot{\delta}_{ij}(t)$ is the total relative velocity between both the buildings; $\bar{\beta}$ is the impact stiffness parameter while $\bar{c}_{ij}(t)$ is the impact element's damping; $\bar{\xi}$ is the damping ratio related to a coefficient of restitution, *e*, which accounts for energy dissipation during the impact (Jankowski 2008). On the other hand, the pounding forces in the transverse direction $F_{yij}^p(t)$, in the vertical direction, $F_{zij}^p(t)$, are calculated by the Coulomb friction model in Eqs. 2.11a–c and 2.12a–e (Chopra 1995, Wriggers 2006a).

$$F_{yij}^P(t) = 0 \text{ for } \delta_{ij}(t) \leq 0$$

$$F_{yij}^P(t) = -\mu_f F_{xij}^P(t) \text{ for } \delta_{ij}(t) > 0 \text{ and } \dot{y}_i(t) - \dot{y}_j(t) > 0$$

$$F_{yij}^P(t) = +\mu_f F_{xij}^P(t) \text{ for } \delta_{ij}(t) > 0 \text{ and } \dot{y}_i(t) - \dot{y}_j(t) < 0 \qquad (2.11a\text{–}c)$$

$$F_{yij}^P(t) = 0 \text{ for } \delta_{ij}(t) > 0 \text{ and } \dot{y}_i(t) - \dot{y}_j(t) = 0$$

$$F_{zij}^P(t) = 0 \text{ for } \delta_{ij}(t) \leq 0$$

$$F_{zij}^P(t) = -\mu_f F_{xij}^P(t) \text{ for } \delta_{ij}(t) > 0 \text{ and } \dot{z}_i(t) - \dot{z}_j(t) > 0$$

$$F_{zij}^P(t) = +\mu_f F_{xij}^P(t) \text{ for } \delta_{ij}(t) > 0 \text{ and } \dot{z}_i(t) - \dot{z}_j(t) < 0 \qquad (2.12a\text{–}e)$$

$$F_{zij}^P(t) = 0 \text{ for } \delta_{ij}(t) > 0 \text{ and } \dot{z}_i(t) - \dot{z}_j(t) = 0$$

Here μ_f is the friction coefficient during pounding. Based on the sliding bearings analysed by Mokha et al. (1990), a sliding base-isolation system is provided by using Teflon sliding bearings between each superstructure and its foundation and consists of Teflon-steel interfaces.

$$F_x(t) = \begin{bmatrix} F_{x1}(t) \\ F_{x2}(t) \\ F_{x3}(t) \\ F_{x4}(t) \\ F_{x5}(t) \\ F_{x6}(t) \\ F_{x7}(t) \\ a_1\left(-\text{sign}\left(\dot{x}_{B1}(t)\right)F_{f1}\right) + a_1\left(F_{x1}(t) + F_{x2}(t) + F_{x3}(t) + F_{x4}(t)\right) - a_1\left(m_1 + m_2 + m_3 + m_4\right)\ddot{x}_g(t) \\ a_2\left(-\text{sign}\left(\dot{x}_{B2}(t)\right)F_{f2}\right) + a_2\left(F_{x5}(t) + F_{x6}(t) + F_{x7}(t)\right) - a_2\left(m_5 + m_6 + m_7\right)\ddot{x}_g(t) \end{bmatrix}$$

$$F_y(t) = \begin{bmatrix} F_{y1}(t) \\ F_{y2}(t) \\ F_{y3}(t) \\ F_{y4}(t) \\ F_{y5}(t) \\ F_{y6}(t) \\ F_{y7}(t) \\ a_1\left(-\text{sign}\left(\dot{x}_{B1}(t)\right)F_{f1}\right) + a_1\left(F_{y1}(t) + F_{y2}(t) + F_{y3}(t) + F_{y4}(t)\right) - a_1\left(m_1 + m_2 + m_3 + m_4\right)\ddot{y}_g(t) \\ a_2\left(-\text{sign}\left(\dot{x}_{B2}(t)\right)F_{f2}\right) + a_2\left(F_{y5}(t) + F_{y6}(t) + F_{y7}(t)\right) - a_2\left(m_5 + m_6 + m_7\right)\ddot{y}_g(t) \end{bmatrix} \qquad (2.13a\text{–}c)$$

$$F_z(t) = \begin{bmatrix} F_{z1}(t) \\ F_{z2}(t) \\ F_{z3}(t) \\ F_{z4}(t) \\ F_{z5}(t) \\ F_{z6}(t) \\ F_{z7}(t) \\ 0 \\ 0 \end{bmatrix}$$

Here $F_{xi}(t)$ ($i = 1,\ldots, 7$) are the wind forces in the longitudinal direction in Eqs. 2.13 a–c; $F_{yi}(t)$ in the transverse direction is similar to $F_{xi}(t)$; and F_{fi} ($i = 1, 2$) is the limiting value of the frictional force during sliding. The direction of friction force will depend on the velocity of the base of structure with respect to the foundation and will be in the opposite direction; sign $(\dot{x}_{Bi})$ ($i = 1, 2$) is the signum function which is used to establish the particular direction of friction force. The value of the frictional force during sliding is expressed as shown in Eq. 2.14.

$$F_{fi} = \mu u_i \left(m_{Bi} + \sum_{j=1}^{4} m_j \right) g \quad (i = 1, 2) \tag{2.14}$$

Here μu_i is the friction coefficient of the sliding bearing and g is acceleration due to gravity.

2.5 Equation of Motion of the Soil-Structure System for Adjacent Buildings

A simplified model for the SSI problem was used considering the pounding force on the behaviour of adjacent buildings. The interaction forces at the soil-structure interface are simulated, using a frequency-independent spring and dashpot set in parallel (Richart et al. 1970). The rectangular dimensions of foundations for both the buildings are converted as circular footings in order to adopt the frequency-independent spring and dashpot set. The simplified model of N- and S-storey coupled buildings resting on the surface of an elastic half-space is shown in Fig. 2.3.

Here $i = (1,2,.,N)$ and $j = (1,2,.,S)$, m_i, k_{xi}, c_{xi}, k_{yi}, c_{yi}, I_{xi} and I_{yi} are the mass, the elastic structural stiffness, damping coefficients and moments of inertia of the related floor about the axes through the centre of mass (CM) and parallel to the x and y axes for Building A and Building B, respectively. The subscripts i and j in Fig. 2.3 are the storey number of the buildings that denote 1, 2,., N for Building A and 1, 2,., S for Building B. Moreover, either subscripts or superscripts of a and b symbolise Building A and Building B, respectively.

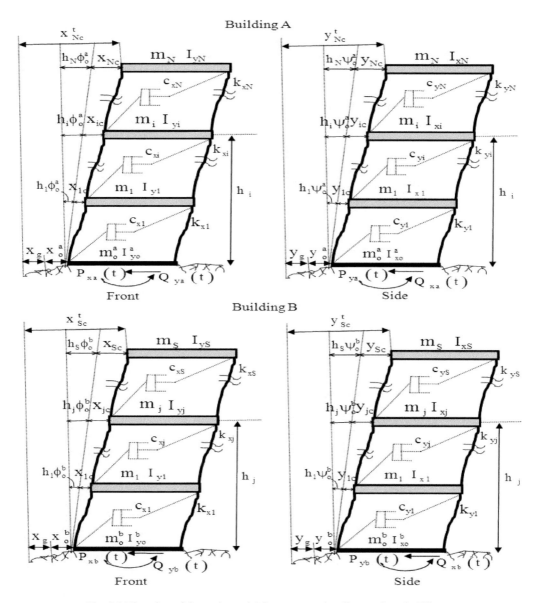

Fig. 2.3 Elevation of dynamic model for asymmetric adjacent shear buildings.

For the horizontal component of ground motion assumed to be uniform over the base of the buildings, the total number of degrees of freedom is $3N + 5$ for N-storey building and therefore $3N + 5$ and $3S + 5$ equations are required for both Building A and Building B, respectively. Equations for Building B are same as shown below for Building A associated with the number of storey (j) of Building B. Hence, only the equations of Building A are expressed here. With reference to Fig. 2.4, translation in the longitudinal (x) and transverse

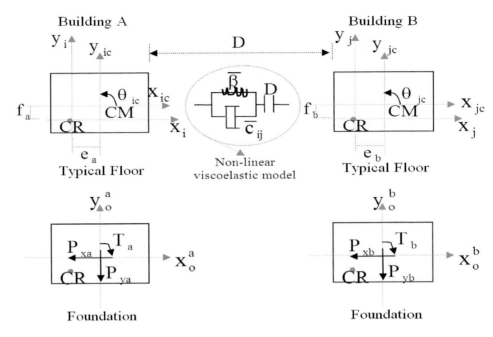

Fig. 2.4 Plan of two-way asymmetric adjacent shear buildings.

(*y*) directions and rotation about the CM of these equations may be expressed as shown in Eqs. 2.15a–c.

As an example, for *N*-storey Building *A*, the total numbers of degrees of freedom, $3N + 5$ are obtained as $3N$ equations of dynamic equilibrium of each floor of the superstructure for translation in x and y directions and rotation about the centre of mass and 5 degrees of freedom due to interaction at the foundation. The $3N$ equations of dynamic equilibrium of each floor of Building *A* may be expressed as

$$[M_a]\{\ddot{x}_{ic}^t\} + [C_{ax}]\{\dot{x}_i\} + [K_{ax}]\{x_i\} + \left[F_{xij}^p(t)\right] = \{0\}$$

$$[M_a]\{\ddot{y}_{ic}^t\} + [C_{ay}]\{\dot{y}_i\} + [K_{ay}]\{y_i\} + \left[F_{yij}^p(t)\right] = \{0\} \qquad (2.15a\text{–}c)$$

$$r_a^2[M_a]\{\ddot{\theta}_i^t\} + f_a[C_{ax}]\{\dot{x}_i\} - e_a[C_{ay}]\{\dot{y}_i\} + [C_{\theta R}^a]\{\dot{\theta}_{ic}\} + f_a[K_{ax}]\{x_i\}$$
$$-e_a[K_{ay}]\{y_i\} + [K_{\theta R}^a]\{\theta_{ic}\} + \left[F_{\theta ij}^p(t)\right] = \{0\}$$

where M_a, C_{ax}, K_{ax}, C_{ay}, and K_{ay} are the $N \times N$ sub-matrices of mass, damping and lateral stiffness in x and y directions of Building *A*, respectively; x_{ic}^t, y_{ic}^t, and θ_i^t are the total displacements of the centre of mass of the floors in the longitudinal and transverse directions, and the total twist of the floors about the vertical z axis in Building *A*, respectively; $F^p_{xij}(t)$ denotes the pounding forces in x direction with the help of the non-linear viscoelastic model (Jankowski 2006, Jankowski 2008); $F^p_{yij}(t)$ and $F^p_{\theta ij}(t)$ and are considered by the Coulomb friction model (Chopra 1995, Wriggers 2006b) and x_i, y_i, and θ_{ic} are the displacement vectors with respect to the base in x and y directions of the centre of resistance (*CR*) and the twist of the floors with respect to the base.

Moreover, two-way asymmetric buildings are modelled with *CR* not being coincident with *CM* along the two horizontal plane axes. The static eccentricities of the centre of resistance from the centre of mass (*e* and *f*) in the x and y axes are same for each floor deck, although the *CR* may vary from storey to storey. Hence, the CR associated with adjacent buildings is assumed to lie at eccentricities e_a, f_a for Building *A* and e_b, f_b for Building *B*. The radii of gyration (r_a and r_b) of any rigid floor deck are about the centre of mass for each building; $K^a_{\theta R}$ and $K^a_{\theta M}$ in Eqs. 2.15a–c and 2.16a–d are the torsional stiffness matrix defined about the *CR* and *CM*, respectively. Furthermore, C_{ax}, C_{ay} and $C^a_{\theta R}$, in Eqs. 2.15a–c are the damping matrices for Building *A*, assumed to be proportional to the stiffness matrices as defined in Eqs. 2.16a–d (Clough and Penzien 1993).

$$\left[K^a_{\theta M}\right] = \left[K^a_{\theta R}\right] + e^2_a\left[K_{ay}\right] + f^2_a\left[K_{ax}\right]$$

$$\left[C_{ax}\right] = \alpha\left[K_{ax}\right], \ \left[C_{ay}\right] = \alpha\left[K_{ay}\right], \ \left[C^a_{\theta R}\right] = \alpha\left[K^a_{\theta R}\right]$$

(2.16a–d)

where α is a constant value in terms of the ratio of the coefficient and stiffness of the buildings. The dashpot constants for adjacent buildings using this damping can be written and simplified as shown for each building in Eqs. 2.17a–b

$$\xi_{ax} = \frac{\alpha\omega_{ax}}{2}, \ \xi_{ay} = \frac{\alpha\omega_{ay}}{2}$$

(2.17a–b)

and where ξ_{ax}, ξ_{ay}, ω_{ax} and ω_{ay} denote the damping ratios of Building *A* and the natural frequencies in the longitudinal and transverse directions, respectively. By assuming 5 per cent damping in the first mode, a damping ratio in the second mode can be found in Eqs. 2.17a–b. The displacement vectors in related directions of Building *A* without SSI effects can be defined by Eqs. 2.18a–f:

$$\{x_{ic}\} = \{x_i\} - f_a\{\theta_{ic}\}, \{\dot{x}_{ic}\} = \{\dot{x}_i\} - f_a\{\dot{\theta}_{ic}\}, \{\ddot{x}_{ic}\} = \{\ddot{x}_i\} - f_a\{\ddot{\theta}_{ic}\}$$

$$\{y_{ic}\} = \{y_i\} + e_a\{\theta_{ic}\}, \{\dot{y}_{ic}\} = \{\dot{y}_i\} + e_a\{\dot{\theta}_{ic}\}, \{\ddot{y}_{ic}\} = \{\ddot{y}_i\} + e_a\{\ddot{\theta}_{ic}\}$$

(2.18a–f)

where x_{ic} and y_{ic} are displacement vectors of degrees of freedom of superstructure about the *CM*. The x^t_{ic}, y^t_{ic} and θ^t_{ic} can be expressed in view of the following relationship of x'_{ic} and y'_{ic} vectors, which are the degrees of freedom of the superstructure defined as in Eqs. 2.19a–e:

$$\left\{x_{ic}^{t}\right\}=x_{o}^{a}\{1\}+x_{g}\{1\}+\phi_{o}^{a}\{h_{i}\}+\{x_{ic}\},\ \left\{x_{ic}^{'}\right\}=x_{o}^{a}\{1\}+\phi_{o}^{a}\{h_{i}\}+\{x_{ic}\}$$

$$\left\{y_{ic}^{t}\right\}=y_{o}^{a}\{1\}+y_{g}\{1\}+\psi_{o}^{a}\{h_{i}\}+\{y_{ic}\},\ \left\{y_{ic}^{'}\right\}=y_{o}^{a}\{1\}+\psi_{o}^{a}\{h_{i}\}+\{y_{ic}\} \qquad (2.19a–e)$$

$$\left\{\theta_{i}^{t}\right\}=\theta_{o}^{a}\{1\}+\{\theta_{ic}\}$$

where x_{o}^{a}, y_{o}^{a}, Ψ_{o}^{a} and ϕ_{o}^{a} are the degrees of freedom at the base associated with translations and rocking about the x and y axes, respectively; θ_{o}^{a} is the twist about the z axis. After substituting Eqs. 2.16a–d, 2.18a–f and 2.19a–e and rearranging into Eqs. 2.15a–c, a more concise form for the $3N \times 3N$ sub-matrices of the superstructure resting on a rigid base of the left upper corner of M^{a}, C^{a} and K^{a} can be written as matrices.

2.6 Interaction Forces

With reference to Fig. 2.3, the equation of motion for the whole foundation system for Building A can be written for translation in the x and y axes, twist about z axis and rocking about the x and y axes, respectively as shown in Eqs. 2.20a–e (Richart et al. 1970).

$$m_{o}^{a}\left(\ddot{x}_{g}+\ddot{x}_{o}^{a}\right)+\{1\}^{T}[M_{a}]\left\{\ddot{x}_{ic}^{t}\right\}+P_{xa}(t)=0$$

$$m_{o}^{a}\left(\ddot{y}_{g}+\ddot{y}_{o}^{a}\right)+\{1\}^{T}[M_{a}]\left\{\ddot{y}_{ic}^{t}\right\}+P_{ya}(t)=0$$

$$r_{a}^{2}m_{o}^{a}\ddot{\theta}_{o}^{a}+r_{a}^{2}\{1\}^{T}[M_{a}]\left\{\ddot{\theta}_{i}^{t}\right\}+T_{a}(t)=0 \qquad (2.20a–e)$$

$$\sum_{i=0}^{N}I_{xi}\ddot{\psi}_{o}^{a}+\{h_{i}\}^{T}[M_{a}]\left\{\ddot{y}_{ic}^{t}\right\}+Q_{xa}(t)=0$$

$$\sum_{i=0}^{N}I_{yi}\ddot{\phi}_{o}^{a}+\{h_{i}\}^{T}[M_{a}]\left\{\ddot{x}_{ic}^{t}\right\}+Q_{ya}(t)=0$$

where I_{xi} and I_{yi} are moments of inertia of the ith floor about the axis through the CM and parallel to the longitudinal and transverse directions, respectively. The m_{o}^{a} is the mass of the foundation of Building A. The h_{i} and h_{j} are the column vector composed of the storey heights of Building A and Building B throughout the foundation to each floor, respectively. Earthquake ground accelerations in the x and y directions are shown as $\ddot{x}_{g}$ and $\ddot{y}_{g}$, respectively. The $P_{xa}(t)$, $P_{ya}(t)$, $T_{a}(t)$, $Q_{xa}(t)$ and $Q_{ya}(t)$ are the interaction forces of Building A based on frequency-independent soil springs and dashpots as shown in Eqs. 2.21a–e (Balendra et al. 1983).

$$P_{xa}(t)=C_{T}\dot{x}_{o}^{a}+K_{T}x_{o}^{a}$$

$$P_{ya}(t)=C_{T}\dot{y}_{o}^{a}+K_{T}y_{o}^{a}$$

$$T_{a}(t)=C_{\theta}\dot{\theta}_{o}^{a}+K_{\theta}\theta_{o}^{a} \qquad (2.21a–e)$$

$$Q_{xa}(t)=C_{\psi}\dot{\psi}_{o}^{a}+K_{\psi}\psi_{o}^{a}$$

$$Q_{ya}(t)=C_{\phi}\dot{\phi}_{o}^{a}+K_{\phi}\phi_{o}^{a}$$

Table 2.1 Spring and Dashpot Constants for Rigid Circular Footing Resting on Elastic Half-space (Richart et al. 1970).

	Sliding	Torsion	Rocking
Spring Coefficient	$K_T = \dfrac{32(1-\upsilon)Gr_o}{7-8\upsilon}$	$K_\theta = \dfrac{16Gr_o^3}{3}$	$K_{\psi,\phi} = \dfrac{8Gr_o^3}{3(1-\upsilon)}$
Mass Ratio	$B_T = \dfrac{(7-8\upsilon)M_T}{32(1-\upsilon)\rho r_o^3}$	$B_\theta = \dfrac{I_\theta}{\rho r_o^5}$	$B_{\psi,\phi} = \dfrac{3(1-\upsilon)I_{\psi,\phi}}{8\rho r_o^5}$
Damping Ratio	$D_T = \dfrac{0.288}{\sqrt{B_T}}$	$D_\theta = \dfrac{0.5}{1+2B_\theta}$	$D_{\psi,\phi} = \dfrac{0.15}{\left(1+B_{\psi,\phi}\right)\sqrt{B_{\psi,\phi}}}$
Coefficient	$C_T = 2D_T\sqrt{K_T M_T}$	$C_\theta = 2D_\theta\sqrt{K_\theta I_\theta}$	$C_{\psi,\phi} = 2D_{\psi,\phi}\sqrt{K_{\psi,\phi}I_{\psi,\phi}}$

where K_T, K_θ, K_ψ, K_ϕ, C_T, C_θ, C_ψ and C_ϕ are the spring and dashpot coefficients of translations about both the *x* and *y* directions, torsion and rocking movements about the *x* and *y* directions, respectively. To begin with, assume that the two buildings remain in the linear elastic range and hence do not yield under earthquake excitation. The definitions of spring and dashpot constants of the static impedance functions are clearly presented in Table 2.1 with various subscripts (Richart et al. 1970).

where M_T, I_θ and $I_{\psi\phi}$ are total mass, polar moment of inertia and moment of inertia of the rigid body for rocking, respectively; G, ρ, υ and v_s are the shear modulus, mass density of half space, Poisson's ratio and shear velocity of the elastic medium, respectively ($G = v_s^2\rho$); and r_o is the radius of the massless disc on the surface of an elastic homogeneous half-space.

2.7 Rigorous Method

In such a case, the equation of motion for couple buildings with whole interactions, such as the SSI, torsional coupling and the pounding involved responses of adjacent buildings modelled with elastic systems at each floor level, is as:

$$\begin{bmatrix} M^a & 0 \\ 0 & M^b \end{bmatrix}\begin{Bmatrix} \ddot{U}^a(t) \\ \ddot{U}^b(t) \end{Bmatrix} + \begin{bmatrix} C^a & 0 \\ 0 & C^b \end{bmatrix}\begin{Bmatrix} \dot{U}^a(t) \\ \dot{U}^b(t) \end{Bmatrix} + \begin{bmatrix} K^a & 0 \\ 0 & K^b \end{bmatrix}\begin{Bmatrix} U^a(t) \\ U^b(t) \end{Bmatrix} + \begin{Bmatrix} F^P(t) \\ -F^P(t) \end{Bmatrix} = -\begin{Bmatrix} P^a(t) \\ P^b(t) \end{Bmatrix} \quad (2.22)$$

where M^a, C^a, K^a, M^b, C^b and K^b are the $(3N + 5) \times (3N + 5)$ dimensional of the mass, damping and stiffness matrices of adjacent buildings, respectively. Moreover, $F^P(t)$, $P^a(t)$ and $P^b(t)$ are vectors containing the forces due to impact between floors with masses m_i,

m_j and loading of the adjacent buildings in that order (see Eqs. (2.23), (2.24a–b) and (2.25a–c)); $\ddot{U}^a(t)$, $\dot{U}^a(t)$, $U^a(t)$, $\ddot{U}^b(t)$, $\dot{U}^b(t)$ and $U^b(t)$ and are the vectors of acceleration, velocity and displacement of the system respectively.

$$
M^a =
\begin{bmatrix}
[M_a] & 0_{N\times N} & 0_{N\times N} & 0 & 0 & 0 & 0 & 0 \\
0_{N\times N} & [M_a] & 0_{N\times N} & 0 & 0 & 0 & 0 & 0 \\
0_{N\times N} & 0_{N\times N} & r_a^2[M_a] & 0 & 0 & 0 & 0 & 0 \\
\{1\}^T[M_a] & \{0\}^T & \{0\}^T & m_o^a & 0 & 0 & 0 & 0 \\
\{0\}^T & \{1\}^T[M_a] & \{0\}^T & 0 & m_o^a & 0 & 0 & 0 \\
\{0\}^T & \{0\}^T & r_a^2\{1\}^T[M_a] & 0 & 0 & r_a^2 m_o^a & 0 & 0 \\
\{0\}^T & \{h_i\}^T[M_a] & \{0\}^T & 0 & 0 & 0 & \sum_{i=0}^{N} I_{xi} & 0 \\
\{h_i\}^T[M_a] & \{0\}^T & \{0\}^T & 0 & 0 & 0 & 0 & \sum_{i=0}^{N} I_{yi}
\end{bmatrix}
\tag{2.23}
$$

where **0** and 1 are the $N \times 1$ column vectors whose elements are equal to zero and one, respectively.

$$
K^a =
\begin{bmatrix}
[K_{ax}] & 0_{N\times N} & f_a[K_{ax}] & -[K_{ax}]\{1\} & 0 & -[K_{ax}]\{f_a\} & 0 & -[K_{ax}]\{h_i\} \\
 & [K_{ay}] & -e_a[K_{ay}] & 0 & -[K_{ay}]\{1\} & +[K_{ay}]\{e_a\} & -[K_{ay}]\{h_i\} & 0 \\
\vdots & & [K_{\theta M}] & -[K_{ax}]\{f_a\} & +[K_{ay}]\{e_a\} & -[K_{\theta M}]\{1\} & +e_a[K_{ay}]\{h_i\} & -f_a[K_{ax}]\{h_i\} \\
 & & \ddots & K_T+\{1\}^T[K_{ax}]\{1\} & 0 & \{f_a\}^T[K_{ax}]\{1\} & 0 & \{1\}^T[K_{ax}]\{h_i\} \\
\vdots & & & & K_T+\{1\}^T[K_{ay}]\{1\} & \{e_a\}^T[K_{ay}]\{1\} & \{1\}^T[K_{ay}]\{h_i\} & 0 \\
 & \text{symm.} & \ddots & & & K_\theta+\{1\}^T[K_{\theta M}]\{1\} & -\{e_a\}^T[K_{ay}]\{h_i\} & \{f_a\}^T[K_{ax}]\{h_i\} \\
\vdots & & & \text{symm.} & & & K_\psi+\{h_i\}^T[K_{ay}]\{h_i\} & 0 \\
 \cdots & & \cdots & & \cdots & & & K_\varphi+\{h_i\}^T[K_{ax}]\{h_i\}
\end{bmatrix}
$$

$$(2.24a\text{–}b)$$

$$
C^a =
\begin{bmatrix}
[C_{ax}] & 0_{N\times N} & f_a[C_{ax}] & -[C_{ax}]\{1\} & 0 & -[C_{ax}]\{f_a\} & 0 & -[C_{ax}]\{h_i\} \\
 & [C_{ay}] & -e_a[C_{ay}] & 0 & -[C_{ay}]\{1\} & +[C_{ay}]\{e_a\} & -[C_{ay}]\{h_i\} & 0 \\
\vdots & & [C_{\theta M}] & -[C_{ax}]\{f_a\} & +[C_{ay}]\{e_a\} & -[C_{\theta M}]\{1\} & +e_a[C_{ay}]\{h_i\} & -f_a[C_{ax}]\{h_i\} \\
 & & & C_T+\{1\}^T[C_{ax}]\{1\} & 0 & \{f_a\}^T[C_{ax}]\{1\} & 0 & \{1\}^T[C_{ax}]\{h_i\} \\
\vdots & & \ddots & & C_T+\{1\}^T[C_{ay}]\{1\} & \{e_a\}^T[C_{ay}]\{1\} & \{1\}^T[C_{ay}]\{h_i\} & 0 \\
 & \text{symm.} & & & & C_\theta+\{1\}^T[C_{\theta M}]\{1\} & -\{e_a\}^T[C_{ay}]\{h_i\} & \{f_a\}^T[C_{ax}]\{h_i\} \\
\vdots & & \ddots & & \text{symm.} & & C_\psi+\{h_i\}^T[C_{ay}]\{h_i\} & 0 \\
 \cdots & & \cdots & & \cdots & & & C_\phi+\{h_i\}^T[C_{ax}]\{h_i\}
\end{bmatrix}
$$

$$P^a(t) = \begin{Bmatrix} [M_a]\{1\}\ddot{x}_g \\ [M_a]\{1\}\ddot{y}_g \\ 0 \\ \left(m_o^a + \{1\}^T[M_a]\right)\ddot{x}_g \\ \left(m_o^a + \{1\}^T[M_a]\right)\ddot{y}_g \\ 0 \\ \{h_i\}^T[M_a]\ddot{y}_g \\ \{h_i\}^T[M_a]\ddot{x}_g \end{Bmatrix}, \quad U^a(t) = \begin{Bmatrix} \left[x_{ic}'\right] \\ \left[y_{ic}'\right] \\ \left[\theta_i^t\right] \\ x_o^a \\ y_o^a \\ \theta_o^a \\ \psi_o^a \\ \phi_o^a \end{Bmatrix}, \quad F^P(t) = \begin{Bmatrix} \left[F_{xij}^P(t)\right] \\ \left[F_{yij}^P(t)\right] \\ \left[F_{\theta ij}^P(t)\right] \\ 0 \\ 0 \\ 0 \\ 0 \\ 0 \end{Bmatrix} \quad (2.25a\text{--}c)$$

where $P^a(t)$, $U^a(t)$ and $F^P(t)$ are $(3N + 5) \times 1$ vectors in Eqs. 2.25a–c. If the two buildings are assumed to be inelastic under the considered ground motion, the coupling equation of motion can be expressed in Eq. 2.26. For the sake of completeness, the equations of pounding forces in the following matrices are briefly presented herein:

$$\begin{bmatrix} M^a & 0 \\ 0 & M^b \end{bmatrix}\begin{Bmatrix} \ddot{U}^a(t) \\ \ddot{U}^b(t) \end{Bmatrix} + \begin{bmatrix} C^a & 0 \\ 0 & C^b \end{bmatrix}\begin{Bmatrix} \dot{U}^a(t) \\ \dot{U}^b(t) \end{Bmatrix} + \begin{Bmatrix} F_a(t) \\ F_b(t) \end{Bmatrix} + \begin{Bmatrix} F^P(t) \\ -F^P(t) \end{Bmatrix} = -\begin{Bmatrix} P^a(t) \\ P^b(t) \end{Bmatrix} \quad (2.26)$$

$$F_a(t) = \left\{ \begin{bmatrix} F_{x1}(t)-F_{x2}(t) \\ F_{x2}(t)-F_{xi}(t) \\ F_{xi}(t)-F_{xN}(t) \\ F_{xN}(t) \end{bmatrix}; \begin{bmatrix} F_{y1}(t)-F_{y2}(t) \\ F_{y2}(t)-F_{yi}(t) \\ F_{yi}(t)-F_{yN}(t) \\ F_{yN}(t) \end{bmatrix}; \begin{bmatrix} F_{\theta1}(t)-F_{\theta2}(t) \\ F_{\theta2}(t)-F_{\theta i}(t) \\ F_{\theta i}(t)-F_{\theta N}(t) \\ F_{\theta N}(t) \end{bmatrix} \right.$$
$$\left. ; F_{xo}^a(t); F_{yo}^a(t); F_{\theta o}^a(t); F_{\psi o}^a(t); F_{\phi o}^a(t) \right\}$$

$$(2.27a\text{--}b)$$

$$F_b(t) = \left\{ \begin{bmatrix} F_{x1}(t)-F_{x2}(t) \\ F_{x2}(t)-F_{xj}(t) \\ F_{xj}(t)-F_{xS}(t) \\ F_{xS}(t) \end{bmatrix}; \begin{bmatrix} F_{y1}(t)-F_{y2}(t) \\ F_{y2}(t)-F_{yj}(t) \\ F_{yj}(t)-F_{yS}(t) \\ F_{yS}(t) \end{bmatrix}; \begin{bmatrix} F_{\theta1}(t)-F_{\theta2}(t) \\ F_{\theta2}(t)-F_{\theta j}(t) \\ F_{\theta j}(t)-F_{\theta S}(t) \\ F_{\theta S}(t) \end{bmatrix} \right.$$
$$\left. ; F_{xo}^b(t); F_{yo}^b(t); F_{\theta o}^b(t); F_{\psi o}^b(t); F_{\phi o}^b(t) \right\}$$

where $F_a(t)$ and $F_b(t)$ are $(3N + 5) \times 1$ and $(3S + 5) \times 1$ vectors consisting of the system inelastic storey restoring forces for both Building *A* and Building *B*, respectively; $F_{xi} = k_{xi}((x_{ic}' - x_o^a) + f_a$

$(\theta_i^t - \theta_o^a) - h_i\phi_o^a)$, $F_{xj} = k_{xj}((x_{jc}' - x_o^b) + f_b(\theta_j^t - \theta_o^b) - h_j\phi_o^b)$ for the elastic range and when up to the storey yield strengths, $F_{xi}^y(t)$, $F_{xj}^y(t)$ are reached; $F_{xi} = \pm F_{xi}^y(t)$ and $F_{xj} = \pm F_{xj}^y(t)$ for the plastic range. For simulating the pounding force during impact F_{xij}^p, $F_{\theta ij}^p$ ($i = 1,2, ..N$; $j = 1, 2, ..S$), the nonlinear viscoelastic model is used between the storey levels of the two adjacent buildings based on the following formula in Eqs. 2.10a–d as both approach period and restitution period of collisions (Jankowski 2006, Mahmoud and Jankowski 2009, Hadi and Uz 2010b, Jankowski 2010):

$$\delta_{ij}(t) = x_{ic}'(t) - x_{jc}'(t) - D , \quad \dot{\delta}_{ij}(t) = \dot{x}_{ic}'(t) - \dot{x}_{jc}'(t) \qquad (2.28a–b)$$

where $\delta_{ij}(t)$ and $\dot{\delta}_{ij}(t)$ in Eqs. 2.28a–b are the total relative displacement and velocity between both buildings with respect to the foundation, respectively. On the other hand, the pounding forces in the transverse direction F_{yij}^p are calculated by the Coulomb friction model in Eqs. 2.11a–c.

2.8 Multi-Degrees of Freedom Modal Equations of Motion

The right-hand side of Eq. 2.22 modified to show the spatial distribution of the effective forces over both the buildings can be represented based on the studies by Chopra and Goel (2004) and Jui-Liang et al. (2009) as follows:

$$P^a(t) = \sum_{n=1}^{3N+5} s_n^a \left(\Gamma_{xn}^a \ddot{x}_g + \Gamma_{yn}^a \ddot{y}_g \right), \quad P^b(t) = \sum_{n=1}^{3S+5} s_n^b \left(\Gamma_{xn}^b \ddot{x}_g + \Gamma_{yn}^b \ddot{y}_g \right) \qquad (2.29a–b)$$

where s_n^a and s_n^b are vectors of the nth modal inertia force distribution equivalent to $M^a\varphi_n^a$ and $M^b\varphi_n^b$ respectively. The φ_n^a is the nth un-damped mode shape obtained from K^a and M^a. The φ_n^b is also obtained the same way as φ_n^a; Γ_{xn}^a, Γ_{yn}^a, Γ_{xn}^b and Γ_{yn}^b in the longitudinal and transverse directions of each building are nth modal participation factors, respectively. The subscript n stands for the nth mode. The subscripts x, x_o, y, y_o, θ, θ_o, Ψ_o and ϕ_o denote sub-vectors relating to translations in both directions, rotations of both superstructure and foundation and rocking degree of freedoms (DOFs) in both the directions, respectively. The nth modal participation factor equals

$$\Gamma_{xn}^a = \frac{\varphi_n^{a^T} \times M^a \times \left[\mathbf{1^T 0^T 0^T} 1\ 0\ 0\ 0\ 0 \right]^T}{\varphi_n^{a^T} \times M^a \times \varphi_n^a}, \quad \Gamma_{yn}^a = \frac{\varphi_n^{a^T} \times M^a \times \left[\mathbf{0^T 1^T 0^T} 0\ 1\ 0\ 0\ 0 \right]^T}{\varphi_n^{a^T} \times M^a \times \varphi_n^a}$$

$$\Gamma_{xn}^b = \frac{\varphi_n^{b^T} \times M^b \times \left[\mathbf{1^T 0^T 0^T} 1\ 0\ 0\ 0\ 0 \right]^T}{\varphi_n^{b^T} \times M^b \times \varphi_n^b}, \qquad (2.30a–d)$$

$$\Gamma_{yn}^b = \frac{\varphi_n^{b^T} \times M^b \times \left[\mathbf{0^T 1^T 0^T} 0\ 1\ 0\ 0\ 0 \right]^T}{\varphi_n^{b^T} \times M^b \times \varphi_n^b}$$

where $\mathbf{1}$ and $\mathbf{0}$ are N × 1 column vectors with all elements equal to one and zero, respectively. Eqs. 2.30a–d show evidently that the nth modal participation factors depend on the direction of the horizontal seismic ground motions. For the 1940 El Centro earthquake motion, this book assumed that only the nth modal displacement responses of the whole system of each building, U_n^a and U_n^b will be excited as defined in Eqs. 2.29a–b by the time variation of $\ddot{x}_g(t)$ and $\ddot{y}_g(t)$. The vertical ground motion is not considered in the modal analysis. The force distribution can be expanded as a summation of modal inertia force distributions as shown in Eqs. 2.29a–b. Hence, Eq. 2.22 can be rewritten as follows:

$$
\begin{aligned}
& \begin{bmatrix} M^a & 0 \\ 0 & M^b \end{bmatrix} \begin{bmatrix} \ddot{U}_n^a(t) \\ \ddot{U}_n^b(t) \end{bmatrix} + \begin{bmatrix} C^a & 0 \\ 0 & C^b \end{bmatrix} \begin{bmatrix} \dot{U}_n^a(t) \\ \dot{U}_n^b(t) \end{bmatrix} + \begin{bmatrix} K^a & 0 \\ 0 & K^b \end{bmatrix} \begin{bmatrix} U_n^a(t) \\ U_n^b(t) \end{bmatrix} + \begin{Bmatrix} F_n^P(t) \\ -F_n^P(t) \end{Bmatrix} \\
& = - \begin{Bmatrix} s_n^a \left(\Gamma_{xn}^a \ddot{x}_g + \Gamma_{yn}^a \ddot{y}_g \right) \\ s_n^b \left(\Gamma_{xn}^b \ddot{x}_g + \Gamma_{yn}^b \ddot{y}_g \right) \end{Bmatrix}
\end{aligned}
\tag{2.31}
$$

where U_n^a and U_n^b are the nth modal displacement response of Building A and Building B, respectively. The F_n^P in Eq. 2.31 is the nth modal pounding force between adjacent buildings while D_n^a and D_n^b are the nth generalised modal coordinates of both the buildings.

$$
U_n^a = \varphi_n^a \times D_n^a = \begin{bmatrix} \varphi_{xn}^{a^T} & \varphi_{yn}^{a^T} & \varphi_{\theta n}^{a^T} & \phi_{x_o n}^a & \phi_{y_o n}^a & \phi_{\theta_o n}^a & \phi_{\psi_o n}^a & \phi_{\phi_o n}^a \end{bmatrix}^T \times D_n^a
$$

$$
U_n^b = \varphi_n^b \times D_n^b = \begin{bmatrix} \varphi_{xn}^{b^T} & \varphi_{yn}^{b^T} & \varphi_{\theta n}^{b^T} & \phi_{x_o n}^b & \phi_{y_o n}^b & \phi_{\theta_o n}^b & \phi_{\psi_o n}^b & \phi_{\phi_o n}^b \end{bmatrix}^T \times D_n^b
\tag{2.32a–b}
$$

where φ_{xn}^a, φ_{yn}^a and $\varphi_{\theta n}^a$ are the N × 1 column sub-vectors of the nth natural vibration mode of the superstructure associated with translational in both directions and rotational DOFs, while the mode shapes of the SSI system consisting of five sub-vectors denote $\phi_{x_o n}^a$, $\phi_{y_o n}^a$, $\phi_{\theta_o n}^a$, $\phi_{\psi_o n}^a$ and $\phi_{\phi_o n}^a$ for Building A. The superscripts of a and b of mode shapes in Eqs. 2.32a–b symbolise the sub-vectors of Building A and Building B, respectively. The nth un-damped modal displacement responses, U_n^a and U_n^b cooperated with generalised modal coordinates for both buildings and can be redefined as:

$$
U_n^a = \begin{bmatrix} D_{xn}^a \varphi_{xn}^{a^T} & D_{yn}^a \varphi_{yn}^{a^T} & D_{\theta n}^a \varphi_{\theta n}^{a^T} & D_{x_o n}^a \phi_{x_o n}^a & D_{y_o n}^a \phi_{y_o n}^a & D_{\theta_o n}^a \phi_{\theta_o n}^a & D_{\psi_o n}^a \phi_{\psi_o n}^a & D_{\phi_o n}^a \phi_{\phi_o n}^a \end{bmatrix}^T
$$

$$
U_n^b = \begin{bmatrix} D_{xn}^b \varphi_{xn}^{b^T} & D_{yn}^b \varphi_{yn}^{b^T} & D_{\theta n}^b \varphi_{\theta n}^{b^T} & D_{x_o n}^b \phi_{x_o n}^b & D_{y_o n}^b \phi_{y_o n}^b & D_{\theta_o n}^b \phi_{\theta_o n}^b & D_{\psi_o n}^b \phi_{\psi_o n}^b & D_{\phi_o n}^b \phi_{\phi_o n}^b \end{bmatrix}^T
\tag{2.33a–b}
$$

or

$$U_n^a = \begin{bmatrix} \varphi_{xn}^a & 0 & 0 & 0 & 0 & 0 & 0 & 0 \\ & \varphi_{yn}^a & 0 & 0 & 0 & 0 & 0 & 0 \\ & & \varphi_{\theta n}^a & 0 & 0 & 0 & 0 & 0 \\ & & & \phi_{x_o n}^a & 0 & 0 & 0 & 0 \\ & & & & \phi_{y_o n}^a & 0 & 0 & 0 \\ & \text{Symm.} & & & & \phi_{\theta_o n}^a & 0 & 0 \\ & & & & & & \phi_{\psi_o n}^a & 0 \\ & & & & & & & \phi_{\phi_o n}^a \end{bmatrix} \begin{bmatrix} D_{xn}^a \\ D_{yn}^a \\ D_{\theta n}^a \\ D_{x_o n}^a \\ D_{y_o n}^a \\ D_{\theta_o n}^a \\ D_{\psi_o n}^a \\ D_{\phi_o n}^a \end{bmatrix}_{8\times1} = T_n^a D_n^a$$

$$(2.34\text{a–b})$$

$$U_n^b = \begin{bmatrix} \varphi_{xn}^b & 0 & 0 & 0 & 0 & 0 & 0 & 0 \\ & \varphi_{yn}^b & 0 & 0 & 0 & 0 & 0 & 0 \\ & & \varphi_{\theta n}^b & 0 & 0 & 0 & 0 & 0 \\ & & & \phi_{x_o n}^b & 0 & 0 & 0 & 0 \\ & & & & \phi_{y_o n}^b & 0 & 0 & 0 \\ & \text{Symm.} & & & & \phi_{\theta_o n}^b & 0 & 0 \\ & & & & & & \phi_{\psi_o n}^b & 0 \\ & & & & & & & \phi_{\phi_o n}^b \end{bmatrix} \begin{bmatrix} D_{xn}^b \\ D_{yn}^b \\ D_{\theta n}^b \\ D_{x_o n}^b \\ D_{y_o n}^b \\ D_{\theta_o n}^b \\ D_{\psi_o n}^b \\ D_{\phi_o n}^b \end{bmatrix}_{8\times1} = T_n^b D_n^b$$

At proportionally damped elastic states, the elements of D_n^a and D_n^b are the same (Chopra and Goel 2004, Jui-Liang and Keh-Chyuan 2007, Lin and Tsai 2007, Jui-Liang et al. 2009), i.e. $D_{xn} = D_{yn} = D_{\theta n} = D_{x_o n} = D_{y_o n} = D_{\theta_o n} = D_{\psi_o n} = D_{\phi_o n}$ for each building, Eqs. 2.34a–b are same as the conventional definition of D_n^a and D_n^b in Eqs. 2.32a–b.

$$\begin{Bmatrix} U_n^a(t) \\ U_n^b(t) \end{Bmatrix} = \begin{bmatrix} T_n^a & 0 \\ 0 & T_n^b \end{bmatrix} \begin{bmatrix} D_n^a(t) \\ D_n^b(t) \end{bmatrix}$$

$$(2.35)$$

2.9 Approximate Method

By substituting Eq. 2.35 into Eq. 2.31 and pre-multiplying both sides of Eq. 2.31 by $\begin{bmatrix} T_n^a & 0 \\ 0 & T_n^b \end{bmatrix}^T$, the result becomes

$$\begin{bmatrix} M_n^a & 0 \\ 0 & M_n^b \end{bmatrix} \begin{Bmatrix} \ddot{D}_n^a(t) \\ \ddot{D}_n^b(t) \end{Bmatrix} + \begin{bmatrix} C_n^a & 0 \\ 0 & C_n^b \end{bmatrix} \begin{Bmatrix} \dot{D}_n^a(t) \\ \dot{D}_n^b(t) \end{Bmatrix} + \begin{bmatrix} K_n^a & 0 \\ 0 & K_n^b \end{bmatrix} \begin{Bmatrix} D_n^a(t) \\ D_n^b(t) \end{Bmatrix} + \begin{Bmatrix} F_n^{aP}(t) \\ -F_n^{bP}(t) \end{Bmatrix}$$

$$= -\begin{Bmatrix} M_n^a \iota \left(\Gamma_{xn}^a \ddot{x}_g + \Gamma_{yn}^a \ddot{y}_g \right) \\ M_n^b \iota \left(\Gamma_{xn}^b \ddot{x}_g + \Gamma_{yn}^b \ddot{y}_g \right) \end{Bmatrix} \qquad (2.36)$$

where $M_n^a = T_n^{a^T} M^a T_n^a$, $C_n^a = T_n^{a^T} C^a T_n^a$, $K_n^a = T_n^{a^T} K^a T_n^a$, $M_n^b = T_n^{b^T} M^b T_n^b$, $C_n^b = T_n^{b^T} C^b T_n^b$ and $K_n^b = T_n^{b^T} K^b T_n^b$ are 8×8 matrices; ι is 8×1 column vector with all elements equal to one. If C^a and C^b in both the buildings are proportionally damped, i.e. $C^a = \alpha M^a + \beta K^a$ and $C^b = \alpha M^b + \beta K^b$, the nth modal damping matrices for both buildings can be represented as:

$$C_n^a = T_n^{a^T} \left(\alpha M^a + \beta K^a \right) T_n^a = \alpha M_n^a + \beta K_n^a$$
$$C_n^b = T_n^b \left(\alpha M^b + \beta K^b \right) T_n^b = \alpha M_n^b + \beta K_n^b \qquad (2.37a\text{–}b)$$

Here, α and β are constants determined by the damping ratios of two specific modes. As the original SSI system is non-proportionally damped, i.e. $C^a \neq \alpha M^a + \beta K^a$, $C^b \neq \alpha M^b + \beta K^b$ Eqs. 2.37a–b becomes $C_n^a \neq \alpha M_n^a + \beta K_n^a$ and $C_n^b \neq \alpha M_n^b + \beta K_n^b$ (Jui-Liang and Keh-Chyuan 2007). It implies that $3N + 5$ and $3S + 5$ non-proportionally damped the multi DOF (MDOF) modal equations of motion as shown in Eq. 2.36 in both Building *A* and Building *B* with the elements of D_n^a and D_n^b unequal to each other even in an elastic state will result in a non-proportionally damped system. The actual condition will close in this case.

$$F_{xijn}^p(t) = 0 \text{ for } \delta_{ijxn}(t) \leq 0,$$

$$F_{xijn}^p(t) = \bar{\beta} \left(\delta_{ijxn}(t) \right)^{3/2} + \bar{c}_{ijn}(t) \dot{\delta}_{ijxn}(t) \text{ for } \delta_{ijxn}(t) > 0 \text{ and } \dot{\delta}_{ijxn}(t) > 0, \quad (2.38a\text{–}c)$$

$$F_{xijn}^p(t) = \bar{\beta} \left(\delta_{ijxn}(t) \right)^{3/2} \text{ for } \delta_{ijxn}(t) > 0 \text{ and } \dot{\delta}_{ijxn}(t) \leq 0$$

$$\delta_{ijxn}(t) = \delta_{ijyn}(t) = D_{xn}^a \varphi_{ixn}^a(t) - D_{xn}^b \varphi_{jxn}^b - D,$$

$$\dot{\delta}_{ijxn}(t) = \dot{D}_{xn}^a \varphi_{ixn}^a(t) - \dot{D}_{xn}^b \varphi_{jxn}^b,$$

$$\dot{\delta}_{ijyn}(t) = \dot{D}_{yn}^a \varphi_{iyn}^a(t) - \dot{D}_{yn}^b \varphi_{jyn}^b,$$

$$\delta_{ij\theta n}(t) = D_{\theta n}^a \varphi_{i\theta n}^a(t) f^a - D_{\theta n}^b \varphi_{j\theta n}^b f^b - D, \qquad (2.39a\text{–}f)$$

$$\dot{\delta}_{ij\theta n}(t) = \dot{D}_{\theta n}^a \varphi_{i\theta n}^a(t) f^a - \dot{D}_{\theta n}^b \varphi_{j\theta n}^b f^b,$$

$$F_{ijn}^{aP} = \begin{bmatrix} F_{xijn}^{aP} & F_{yijn}^{aP} & F_{\theta ijn}^{aP} & 0 & 0 & 0 & 0 & 0 \end{bmatrix}_{(3N+5) \times 1}^T$$

where $F^{ap}_{\theta ijn}$ can be calculated as F^{ap}_{xijn} using the related $\delta_{ij\theta n}(t)$ and $\dot{\delta}_{ij\theta n}(t)$; $F^{ap}_n(t) = T^{aT}_n F^{ap}_{ijn}$ in Eq. 2.36 is 8×1 column vector in the calculation of pounding forces. The modal displacement histories of both the buildings, $D^a_n(t)$ and $D^b_n(t)$, are solved using the direct integration method of Eq. 2.36. Hence, the total response histories of the non-proportionally damped adjacent buildings resting on the surface of elastic-half space are obtained as:

$$U^a(t) = \sum_{n=1}^{3N+5} U^a_n(t) \approx \sum_{n=1}^{3N+5} T^a_n D^a_n(t)$$

$$U^b(t) = \sum_{n=1}^{3S+5} U^b_n(t) \approx \sum_{n=1}^{3S+5} T^b_n D^b_n(t)$$

$$(2.40a\text{–}b)$$

To obtain an agreeable output the same as the conventional modal displacement history analysis, only the first three modal responses need to be included in the summation in Eqs. 2.40a–b. This will be shown in the next numerical examples by Jui-Liang et al. (2009) and Sivakumaran and Balendra (1994). Figure 2.5(a) shows the front elevation of the lumped model of the SSI system for adjacent buildings considering the pounding effects. Equation 2.22 can be decomposed into 3N+3S+10 equations, each representing a single DOF (SDOF) modal system in Fig. 2.5(b) by calculating the corresponding modal damping for each vibration mode (Balendra et al. 1983, Novak and Hifnawy 1983b). The MDOF modal equation of motion shown in Eq. (2.36) can be represented by a MDOF modal system resting on the surface of elastic base as shown in Fig. 2.5(c), in contrast to the equivalent SDOF

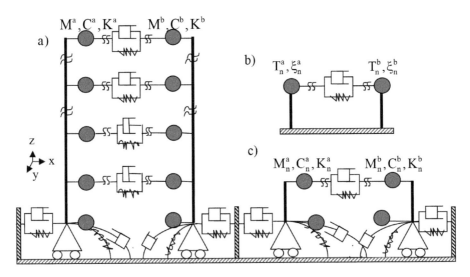

Fig. 2.5 Front elevation of a) the lumped model of the adjacent buildings resting on the surface of elastic base associated with pounding effects, b) the *n*th SDOF modal system for adjacent buildings with corresponding damping ratio and c) the *n*th MDOF modal system resting on an elastic half space for adjacent buildings (Jui-Liang et al. 2009).

modal equation of motion represented by a SDOF modal system. With the corresponding elements of both M^a, C^a, K^a for Building A and M^b, C^b, K^b for Building B, the nth MDOF modal equations of motion, M_n^a, C_n^a, K_n^a, M_n^b, C_n^b, K_n^b can be arranged. Hence, Eqs. 2.23 and 2.25 expressed in a similar form to those of the equations in Appendix I. Moreover, M_n^a, C_n^a, K_n^a are in the same form as that of M_n^b, C_n^b, K_n^b as shown in Appendix I, with superscript of b replaced by superscript of a. These matrices represent the corresponding modal properties of the superstructure and the impedance functions at the foundation of the SSI system of each building.

2.10 Summary

This chapter has the equation of motion for the base-isolated coupled buildings modelled for investigating the pounding effects without the SSI effects. Secondly, to investigate the effects of the SSI system on fixed coupled buildings, the equation of motion of the SSI system including the pounding effects are illustrated in this chapter. In order to make a simple and real-valued modal response history analysis for engineering applications, the proposed approximate method for single buildings used by Jui-Liang et al. (2009) is applied to the modal equations of the motion of the coupled buildings. Chapter 3 shows the modelling of passive and active dampers for seismic response mitigation of a coupled building connected to each other by either passive or active dampers. Before describing control strategies based on GA, non-dominated sorting genetic algorithm and pareto-optimal solutions in the following chapters, the modelling of MR dampers is investigated herein. One of the challenges in the application of MR dampers is use of an appropriate control algorithm to determine the command voltage of the MR damper. Many control algorithms are proposed to control the behaviour of MR dampers or other semiactive devices.

3

Seismic Isolation and Energy-dissipating Devices

3.1 Introduction

In this chapter, control devices are introduced to provide insight into the effect of adjacent buildings on dynamics of the adjacent system. Some concepts of coupled building control in previous research studies noted that the concepts mentioned above are to add damping to the adjacent building system (Klein and Healy 1985, Christenson et al. 1999). After a brief overview of these concept studies, the passive damping devices which are one of the most important damping classes to reduce the response on the dynamics of the system are underscored on coupled building control. The second part of this chapter proves the efficacy of optimal passive and active dampers for achieving the best results in seismic response mitigation of a coupled building connected to each other by either passive or active dampers. Additionally, the studies of many researchers about passive devices are described in this chapter. Their parametric studies are presented for finding the effective passive devices on the dynamic response of damper-connected adjacent buildings under earthquake excitation, using hinged link based on the optimum damper stiffness and coefficient (Xu et al. 1999).

Finally, background material for models and algorithms used in semi-active control systems with MR dampers. Adequate modelling of the control devices is essential for predicting the behaviour of the controlled system. Here, the MR damper is modelled, using a modified Bouc-Wen hysteresis model. In controlling the MR damper, the desired control force cannot be directly commanded because the control force generated by the device is dependent on the local responses of the structure where it is installed. Only the voltage applied to the MR damper can be controlled. In this chapter, the model of the device and the semi-active control algorithm used with the device, the clipped-optimal control algorithm, are discussed. A modified version of this algorithm is also proposed in this chapter.

3.2 Damage Control under Earthquake Loading

The most common systems for structural damage control during earthquake events can be categorised under three aspects: traditional systems, innovative structural control and combination of both of these two. A large variety of energy dissipation devices can be used as pounding mitigation devices. After a brief description of energy dissipation devices in seismic control of building structures, this section reviews the various types of energy dissipation devices. Various types of control devices have been widely used as supplemental damping strategies in order to mitigate the effects of earthquakes and high wind load on civil engineering structures. Several types of dampers have been studied on to structures as paramount interest over the past two decades. These dampers include fluid visco-elastic dampers (Zhang and Xu 1999, Zhang and Xu 2000, Yang et al. 2003, Uz 2009, Uz and Hadi 2009, Zhu et al. 2011), friction dampers (Zhang and Xu 1999, Zhang and Xu 2000, Yang et al. 2003, Zhu et al. 2011), active devices (Bhaskararao and Jangid 2006a, 2006b, Ng and Xu 2006) and semi-active magnetorheological (MR) dampers (Xu and Zhang 2002, Ying et al. 2003).

3.3 Classification of Structural Control Devices

A great number of protective systems for structures have been invented because of the need to provide a safer and more efficient design. Control devices on earthquake zone have been improved since the 1970's. The purpose of structural control is to absorb the energy due to dynamic loadings, such as winds, earthquakes and vehicle loads. Modern structural protective systems can be categorised into three classes as shown graphically in Fig. 3.1.

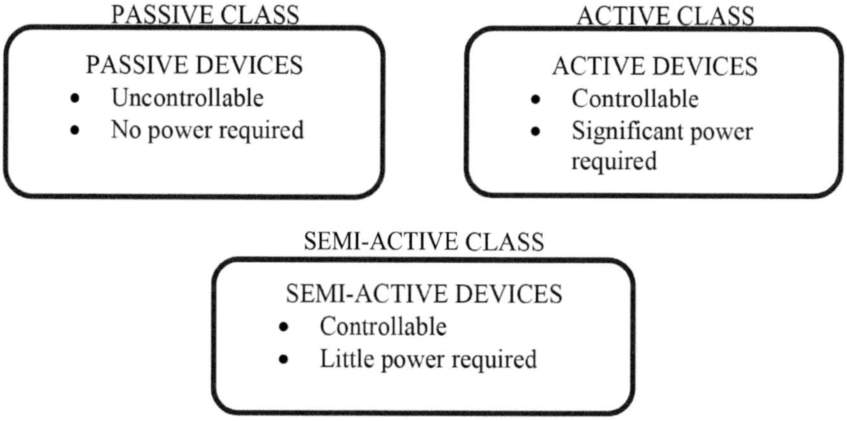

PASSIVE CLASS

PASSIVE DEVICES
- Uncontrollable
- No power required

ACTIVE CLASS

ACTIVE DEVICES
- Controllable
- Significant power required

SEMI-ACTIVE CLASS

SEMI-ACTIVE DEVICES
- Controllable
- Little power required

Fig. 3.1 Control classes and control devices.

The first class of damping devices is passive. They are uncontrollable. The basic function of passive damping devices is to consume a part of the input energy, reduce energy dissipation on structural members and minimise damage on structures. Contrary to, semi-active or active devices, there is no need for an external supply of power. The second class of damping devices is active. The active damping devices are controllable and require significant power. The displacement of structures is controlled or modified by action of the active damping devices through an external supply of power. The third class of damping devices is semi-active. The semi-active damping devices combine the aspect of active and passive damping devices, which involve the amount of external energy to adjust their mechanical properties, unlike fully active systems. The semi-active devices cannot add energy to the structure.

Over the past decade, many conferences have been organised on structural control for civil structures. One of them is the First World Conference on Structural Control that was held in Pasadena, California (Housner et al. 1994). The World Conferences on Structural Control were held in Kyoto, Japan in 1998 and Como, Italy in 2002. These conferences highlight the importance of continued studies on structural control for civil structures. Nowadays, the use of control devices in buildings has become important in order to alter or control the dynamic behaviour of buildings. Additional guidelines and design provisions for energy dissipation systems are provided in NEHRP Commentary on the Guidelines for the Seismic Rehabilitation of Buildings (FEMA 274). An example of each of these control devices is shown in the thesis of Uz (2013). The following sections focus on these systems before providing a detailed review of these control systems.

3.4 Passive Control Systems

Housner et al. (1994) mentioned that passive control devices consume energy which comes from dynamic loadings. Passive control devices are obtained by insertion to the civil structure. All passive control devices have both the stiffness and damping in order to achieve a limitation in the shift of buildings towards the other and to consume the energy. Thereby, passive devices are characterised by their control forces and fixed characteristics of the devices. Another important advantage is the reinstalment of the system after the earthquake for use of the structure. Soong and Dargush (1997) determined their passive control devices include metallic, friction, visco-elastic and viscous fluid dampers, tuned mass dampers and tuned liquid dampers. Passive devices are loaded in terms of protecting the structure from dynamic loading.

One of the most important damping devices in passive control is base isolation. Warnotte et al. (2007) emphasised that base isolation systems cannot be placed either to diminish the individual displacement of one structure or to connect two adjacent structures. Base-isolation systems placed at the foundation of a structure can be used to absorb and reflect some of the earthquake input energy which can be transmitted to the structure (Warnotte et al. 2007). Another passive energy device is the tuned mass damper (TMD). The energy transfer from the primary structure to the TMD by means of the motion of TMD can be placed between story levels in a passive system. Figure 3.2 shows examples of various passive control systems. Passive control devices are popular and are widely employed. Passive devices are quite simple to design and build. However, their performance is sometimes limited. Therefore, for achieving better performance of passive control devices, optimum damper properties are implemented to protect against one particular dynamic loading.

Qi and Chang (1995) described the implementation of viscous dampers that have several inherent and significant advantages including linear viscous behaviour, insensitivity to stroke and output force, easy installation, almost free maintenance, reliability and longevity. Currently, fluid dampers to attain more performance during seismic events has been used by more than 110 major structures. The new Arrowhead Regional Medical Centre in Colton, California, installing 186 dampers and the new fifty-five floor Torre Mayor Office building in Mexico City, Mexico, using 98 dampers are some of these projects (Taylor and Constantinou 1998). Specifications for these dampers are provided in Table 3.1.

The horizontal flexibility to move the fundamental period of the structure away from the ground excitation components can be provided by using base-isolation systems. Using various devices, such as isolators and dissipater dampers (vibration absorbers), passive

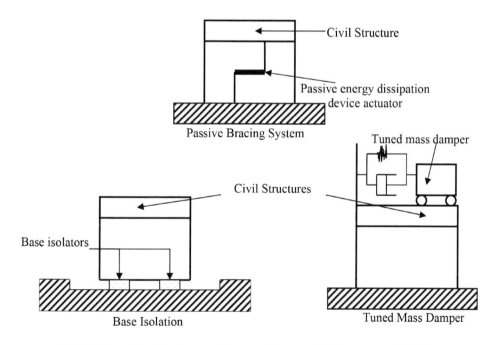

Fig. 3.2 Examples of various passive control systems (Christenson et al. 1999).

Table 3.1 San Bernardino County Medical Centre Damper Specifications (Taylor and Constantinou 1998).

Displacement	=	1.2 m
Max Damping Force	=	145 tonnes
Max Operating Velocity	=	1.6 m/s
Power Dissipation	=	2,170,000 watts
Length	=	4.5 m extend
Diameter	=	0.36 m
Weight	=	1360 kg
Quantity Required	=	186 units

control systems which can deform and yield during external loading help in dissipation of large amounts of input energy. Because of the high energy-absorbing capacities of hysteresis elements, damage to other elements of the building or buildings is reduced. This chapter mainly focusses on base isolation systems and visco-elastic dampers in passive energy dissipation (PED) devices.

3.5 Active Control Systems

One important structural control system is the active control device. Yao (1972) recommended the active control devices for civil structures. These control devices create a force in the structure to counteract the energy of dynamic loading. Thus, different loading conditions and different vibration modes are controlled or accommodated by means of the active control devices (Housner et al. 1994). The feedback from sensors measuring the amplitude of a structure to manage the properties of structural members throughout mechanical actuators is used by active control devices. Figure 3.3 shows some different types of active control devices in use, some of which are as follows: active mass driver, active base isolation and active bracing. A controller (computer) collects records from the sensors to activate devices for amending the structure's amplitude continuously during excitation. Active devices can increase the performance over passive control devices, determining appropriate control forces. For example, a passive tuned mass damper must provide control forces based on the response of the floor. In contrast, active control devices can measure the response using a controller. One problem here is that since the mechanism of these devices depends on external power supply, the latter must not be interrupted during an earthquake, otherwise, the whole system can remain idle at the time when the supply is required.

As a result, active control devices are more complex than passive devices, requiring sensors and controller equipment (Warnotte et al. 2007).

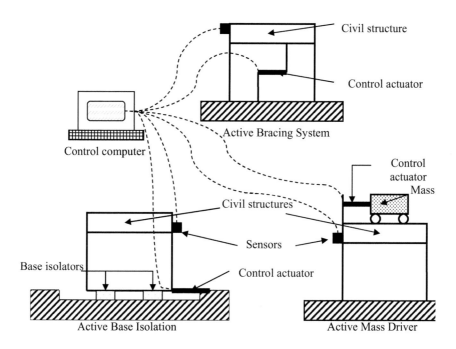

Fig. 3.3 Examples of various active control systems.

3.6 Semi-active Control Systems

Semi-active control systems combine the aspect of active and passive damping devices, which involve the amount of external energy to adjust their mechanical properties, unlike fully active systems. A schematic configuration of the structural control methods described above is shown in Fig. 3.4. When the control actuators do not supply mechanical energy directly to the primary structure, semi-active control systems can absorb only the energy of the input excitation. Once the control actuators supply, the mechanisms of semi-active control systems act as active control systems, directly providing a force to the structure from the control actuator or vise-versa. The control forces are improved in conjunction with adjustment of damping or stiffness characteristics of the semi-active control systems (Spencer Jr et al. 1997, Dyke et al. 1998, Symans and Constantinou 1999, Cheng et al. 2006, Warnotte et al. 2007).

As examples of semi-active control devices, variable-orifice fluid dampers, variable-stiffness control devices, semi-active tuned mass dampers, semi-active tuned liquid dampers, controllable friction dampers, electrorheological dampers, magnetorheological dampers, controllable impact dampers and controllable fluid dampers can be given (Cheng et al. 2008).

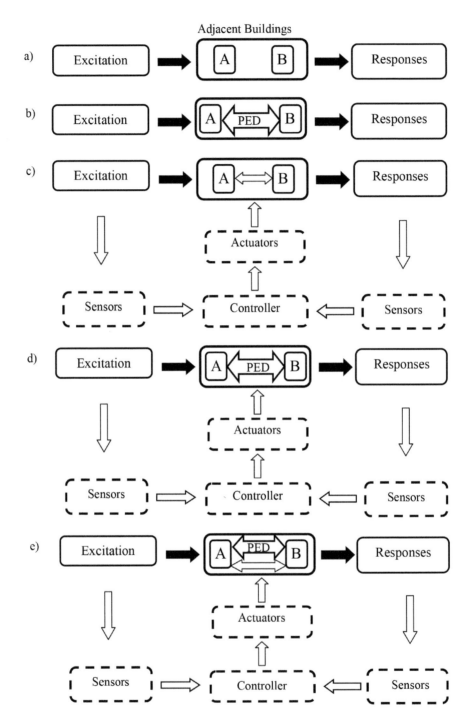

Fig. 3.4 Various control schemes for adjacent buildings, a) conventional design, b) structural control with passive systems, c) with active systems, d) with semi-active systems and e) with hybrid systems.

3.7 Dynamics of Controlled Adjacent Buildings

In this section, two adjacent buildings with n and m stories (n > m) are shown herein, coupled by n_a active control devices with hydraulic actuators and MR dampers. Equations of motion of adjacent buildings are shown in Eqs. (3.1) and (3.2). Adjacent buildings that have flexible columns and mass concentrated at the rigid slabs can be obtained by writing the equilibrium equations from the free body diagram of each of the lumped mass of the building.

- Equation of motion of Building *A*:

$$M_1\ddot{X}_1 + C_1\dot{X}_1 + K_1X_1 = -M_1E_1\ddot{X}_g \tag{3.1}$$

- Equation of motion of Building *B*:

$$M_2\ddot{X}_2 + C_2\dot{X}_2 + K_2X_2 = -M_2E_2\ddot{X}_g \tag{3.2}$$

Equations 3.1 and 3.2 should be solved simultaneously. When the related control is considered, a convenient matrix form can be developed by first combining these equations that lead to the expression

$$\begin{bmatrix} M_1 & 0 \\ 0 & M_2 \end{bmatrix}\begin{Bmatrix} \ddot{X}_1 \\ \ddot{X}_2 \end{Bmatrix} + \left(\begin{bmatrix} C_1 & 0 \\ 0 & C_2 \end{bmatrix} + \begin{bmatrix} c_{d(m,m)} & 0_{(m,s)} & -c_{d(m,m)} \\ 0_{(s,m)} & 0_{(s,s)} & 0_{(s,m)} \\ -c_{d(m,m)} & 0_{(m,s)} & c_{d(m,m)} \end{bmatrix} \right)\begin{Bmatrix} \dot{X}_1 \\ \dot{X}_2 \end{Bmatrix}$$

$$+ \left(\begin{bmatrix} K_1 & 0 \\ 0 & K_2 \end{bmatrix} + \begin{bmatrix} k_{d(m,m)} & 0_{(m,s)} & -k_{d(m,m)} \\ 0_{(s,m)} & 0_{(s,s)} & 0_{(s,m)} \\ -k_{d(m,m)} & 0_{(m,s)} & k_{d(m,m)} \end{bmatrix} \right)\begin{Bmatrix} X_1 \\ X_2 \end{Bmatrix} = \begin{bmatrix} -M_1E_1 \\ -M_2E_2 \end{bmatrix}\ddot{X}_g + \begin{bmatrix} P_1 \\ P_2 \end{bmatrix}F_{mr}(t) \tag{3.3}$$

Equations of motion in Eq. 3.3 which is explained in Appendix I can be transformed into first order state equations. The $c_{d(m,m)}$ and $k_{d(m,m)}$ are (m × m) diagonal matrix of the additional damping and stiffness matrices due to the instillation of the related dampers. The subscript *s* in Eq. 3.3 denotes the (n – m) difference of the number of storeys of both buildings. Note that $F_{mr}(t)$ in Eq. 3.3 denotes the control force for MR dampers while the control force for active dampers is shown as U(t).

3.8 State Space Equations

By defining the state vector, $X = \{ X_1\ X_2\ \dot{X}_1\ \dot{X}_2 \}^T$, noting that Eq. 3.3 may be rewritten as

$$\begin{Bmatrix} \ddot{X}_1 \\ \ddot{X}_2 \end{Bmatrix} = -M^{-1}C\begin{Bmatrix} \dot{X}_1 \\ \dot{X}_2 \end{Bmatrix} - M^{-1}K\begin{Bmatrix} X_1 \\ X_2 \end{Bmatrix} + M^{-1}\begin{Bmatrix} -M_1E_1 \\ -M_2E_2 \end{Bmatrix}\ddot{X}_g + M^{-1}\begin{Bmatrix} P_1 \\ P_2 \end{Bmatrix}F_{mr}(t) \tag{3.4}$$

From Eq. (3.4) the velocity of the state vector can be obtained as

$$
\dot{X} = \left\{ \begin{array}{c} \dot{X}_1 \\ \dot{X}_2 \\ \ddot{X}_1 \\ \ddot{X}_2 \end{array} \right\} = \left[\begin{array}{cc} 0_{(n+m)\times(n+m)} & \mathbf{I}_{(n+m)\times(n+m)} \\ -M^{-1}K & -M^{-1}C \end{array} \right] \left\{ \begin{array}{c} X_1 \\ X_2 \\ \dot{X}_1 \\ \dot{X}_2 \end{array} \right\} \tag{3.5}
$$

$$
+ \left[\begin{array}{c} 0_{(n+m)\times 1} \\ M^{-1}\Gamma \end{array} \right] \ddot{X}_g(t) + \left[\begin{array}{c} 0_{(n+m)\times n_a} \\ M^{-1}\Lambda \end{array} \right] F_{mr}(t)
$$

where

$$
\Lambda = \left[\begin{array}{c} P_1 \\ 0 \\ P_2 \end{array} \right], \Gamma = \left[\begin{array}{c} -M_1 E_1 \\ -M_2 E_2 \end{array} \right], A = \left[\begin{array}{cc} 0_{(n+m)\times(n+m)} & \mathbf{I}_{(n+m)\times(n+m)} \\ -M^{-1}K & -M^{-1}C \end{array} \right]
$$

$$
E = \left[\begin{array}{c} 0_{(n+m)\times 1} \\ M^{-1}\Gamma \end{array} \right], B = \left[\begin{array}{c} 0_{(n+m)\times n_a} \\ M^{-1}\Lambda \end{array} \right] \tag{3.6a–e}
$$

where E_1 and E_2 are $n \times 1$ and $m \times 1$ unity matrices, respectively. The P_1 and P_2 are given in Appendix I. Here, $\mathbf{I}$ is an identity matrix and 0 in Λ matrix is a $(s \times n_a)$ matrix containing zero. The $F_{mr} = [f^1_{mr} \cdots f^i_{mr} f^m_{mr}]^T$ is control input vector. The equation of motion in Eq. 3.5 can be arranged as

$$
\dot{X} = AX + BF_{mr}(t) = E\ddot{X}_g(t) \tag{3.7}
$$

Since only earthquake loading is considered, the equations of motion can be written as

$$
\dot{X} = AX + E\ddot{X}_g(t) \tag{3.8}
$$

Equation (3.8) helps the investigation of the uncontrolled adjacent buildings system in order to understand the efficiency of MR dampers between both buildings.

3.9 Feedback Control

The measurement output y_m in a standard feedback control system usually needs a full state feedback measurement in the form (Arfiadi 2000)

$$
\begin{aligned} x &= C_w X + D_w F_{mr} \\ y_m &= C_m X + D_m F_{mr} + v \end{aligned} \tag{3.9a–b}
$$

In which y_m is the vector of measured outputs, x is the regulated output vector and v is the measurement noise vector. Note that displacements, velocities and absolute accelerations of adjacent buildings can be included to the controlled output defined. By choosing the appropriate entry in the regulation matrix, certain regulated output that needs to be minimized can be imposed. For example, if the regulated output in Eq. 3.10 is taken as the relative displacement of the floors with respect to the ground, the matrix C_w can be chosen as:

$$C_w = \left[I_{(N+M)\times(N+M)} \quad 0_{(N+M)\times(N+M)} \right]$$ (3.10)

For feedback control in Eqs. 3.9a–b, the control force for active control systems ($F_{mr}(t)$ substitutes with U(t) in this case) in Eq. 3.7 can be written by

$$U(t) = -GX$$ (3.11)

Hence, the closed loop system in Eq. 3.7 can be rearranged, using Eq. 3.11

$$\dot{X} = (A - BG)\, X + E\ddot{X}_g(t)$$ (3.12)

Equation 3.12 becomes a form of standard feedback to be used for classical control method. As some cases clearly stated by Meirovitch (1992), the measurement of all the states can be impractical. So that the feedback control having only a certain measurement can be preferable. For this case, two types of output feedback control can be conducted. The first which is given in detail in the thesis of Arfiadi (2000) is known as the observer based controllers or dynamic output feedback controllers by estimating the state from the measurement output in Eqs. 3.9a–b. With the available measurement to be chosen, the control force is regulated as:

$$U(t) = -G\hat{X}$$ (3.13)

where $\hat{X}$ is the observer state vector that estimates the actual state *X*. Herein, without going into a detail description until the discussion of H_2/LQG control algorithm, the observer state vector may be assumed to have the form (Meirovitch 1992)

$$\dot{\hat{X}} = A\hat{X} + BU(t) + L_g\,(y_m - C_m\hat{X})$$ (3.14)

where L_g is the gain matrix for state estimator with the state observer technique, which is determined by solving an algebraic Riccati equation in control toolbox in MATLAB (R2011b). The observer state vector $\hat{X}$ defined in Eq. 3.14 helps to regulate the feedback control system in Eq. 3.13. As can be seen that the system needs an online computation whether the observer state is in good agreement with the actual state or not. For further details about the error dynamic the reader is referred to Arfiadi (2000). In a direct (static) output feedback, without constructing an observer to estimate the actual state, the system can directly utilize the measurement output. When D_w in Eqs. 3.9a–b is considered as zero matrices, the control force in direct output feedback can be obtained by multiplying the measurement with the gain matrix as

$$U(t) = -Gx = -GC_w X \tag{3.15}$$

So that the closed loop system can be taken by substituting Eq. 3.15 into Eq. 3.12 as

$$\dot{X} = (A - BGC_w) X + E\ddot{x}_g \tag{3.16}$$

In output feedback controllers, the regulated output x can be included with the absolute acceleration of the structure based on the requirement of the system.

$$U(t) = -G_d \begin{bmatrix} X_1 \\ X_2 \end{bmatrix} - G_v \begin{bmatrix} \dot{X}_1 \\ \dot{X}_2 \end{bmatrix} - G_a C_{sa} \begin{bmatrix} X_1 \\ X_2 \\ \dot{X}_1 \\ \dot{X}_2 \end{bmatrix} \tag{3.17}$$

where G_d, G_v and G_a are the gain matrices and if there is no feedback from the corresponding measurement output, the element of these gain matrices contains zero (i.e. see Eqs. 3.20a–c).

$$C_{sa} = \begin{bmatrix} -M_{cl}^{-1}K_{cl} & -M_{cl}^{-1}C_{cl} \end{bmatrix}$$
$$M_{cl} = M + \Lambda G_a, K_{cl} = K + \Lambda G_d, C_{cl} = C + \Lambda G_v \tag{3.18a–d}$$

Rearranging Eq. 3.17, the control force can be shown as $U(t) = -G_z X$. Here, X is a state vector in Eq. 3.7.

$$G_z = [G_d \quad G_v] + G_a C_{sa} \tag{3.19}$$

According to the chosen feedback, the gain matrix can be written for example as:

$$G_d = \begin{bmatrix} 0 & 0 & 0 & 0 & 0 & G_{d1} & 0 & 0 & 0 & 0 & 0 & G_{d2} \end{bmatrix}$$
$$G_v = \begin{bmatrix} 0 & 0 & 0 & 0 & 0 & G_{v1} & 0 & 0 & 0 & 0 & 0 & 0 \end{bmatrix} \tag{3.20a–c}$$
$$G_a = \begin{bmatrix} 0 & 0 & 0 & 0 & 0 & 0 & 0 & 0 & 0 & 0 & 0 & G_{a1} \end{bmatrix}$$

where G_{d1}, G_{d2}, G_{v1} and G_{a1} are gains to be determined. The closed loop system can be obtained as

$$\dot{X} = A_{cl}X + E\ddot{x}_g \tag{3.21}$$

where $A_{cl} = A - BG_z$ and the feedback is the top floor of displacement of both buildings, the velocity of top floor of Building *A* and the absolute acceleration of the top floor of Building *B* as shown in Eqs. 3.20a–c. Regulated output C_w can be chosen as inter-storey drifts, the top floor displacement of both buildings and the control force to obtain the optimum controller gains.

3.10 Stability of Active and Other Control Systems

Active control system may cause instability if not designed properly. To avoid the instability of the structure, the simplest way is to ensure that the eigen-values of the closed loop system are placed in the left plane of the s-plane. This constraint is incorporated to the fitness function by simply setting the fitness of the individual having positive real-part eigen-values to a very small positive value that can still be accepted by the computer. If the system has negative real part of the eigen-values then the system is called asymptotically stable (Meirovitch 1992, Michalewicz 1996). In this book, Routh-Hurwitz criterion and the system matrix in the equation of motion of adjacent buildings are used as stability criterion in classical control design. Detailed discussion of these criterion is not the purpose of this research study. To perform Routh-Hurwitz test and the eigenvalue of the closed loop system for direct output feedback control systems, the reader is referred to Arfiadi (2000).

3.11 Modelling of Viscous Dampers

As an energy dissipation device, viscous dampers have been used to diminish earthquake damage to trade structures in many construction projects in current years (Hou 2008). For each displacement degree of freedom, independent damping properties may be specified. The damping properties are based on the Maxwell model of viscous damper having a linear or nonlinear damper in series with a spring. The cyclic response of a fluid viscous device is dependent on the velocity of motion. As recommended by Seleemah and Constantinou (1997) and by seismic design guidelines such as FEMA 273 (BSSC 1997), the linear damper behaviour is given by

$$F_d = C_d \left\{ \begin{matrix} \dot{X}_1 \\ \dot{X}_2 \end{matrix} \right\}^e + K_d \left\{ \begin{matrix} X_1 \\ X_2 \end{matrix} \right\} \tag{3.22}$$

where F_d is total output force provided by the damper, C_d and K_d are the damping coefficient and the spring constant matrices, arranging the velocity across the damper and the displacement across the spring, respectively. The e is the damping exponent. The damping exponent must be positive. The practical range between $e = 0.5$ and 2.0 is determined by Hou (2008) and Tezcan and Uluca (2003). In the numerical data of this book, e is taken as unity. Equation 3.22 consists of two parts. The first is the damping force which is proportional to e; while the second is the restoring force (Hadi and Uz 2009, Uz and Hadi 2009).

3.12 Modelling of Magnetorheological Damper

The magnetorheological (MR) damper is one of the most promising semi-active devices that uses MR fluid in order to provide controllable devices that employ MR fluids (Dyke et al. 1996a, Spencer Jr et al. 1997, Dyke et al. 1998, Yang et al. 2002). The MR fluids that

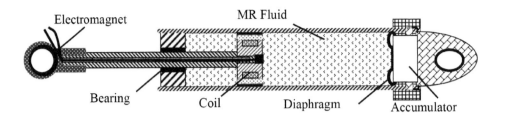

Fig. 3.5 Schematic diagram of MR damper.

were initially discovered by Rabinow (1948) can be the property of a specific class of smart materials with rheological properties controllable rapidly by an applied magnetic field. Figure 3.5 shows the schematic of MR damper. For civil engineering applications, MR dampers are considerably attractive in terms of large force capacity, high stability, robustness and reliability. In addition, they are relatively inexpensive to manufacture and maintain. The MR dampers have also stable hysteretic behaviour over a wide temperature range that makes them suitable for both indoor and outdoor applications.

Because of their mechanical simplicity, high dynamic range and low power requirements (only a battery for power), have been studied by a number of researchers for seismic protection of civil structures (Dyke et al. 1996b, Spencer Jr et al. 1997, Jung et al. 2006, Ok et al. 2007, Shook et al. 2008, Bitaraf et al. 2010, Bitaraf et al. 2012) are considered as good candidates in terms of reducing the structural vibrations. As an example of Dyke (1996), the peak power required is less than 10 watts, which would allow the damper to be operated continuously for more than an hour on a small camera battery. The current for the electromagnet is supplied by a linear current driver of 120 volts AC and generating a 0 to 1 amp current that is proportional to a commanded DC input voltage in the range 0–3 V. As a summary of the design parameters for the large-scale MR damper, Table 3.2 is given:

Forces of up to 3000 N can be generated. To summarise, the following three types of dynamic models are given herein. In this book, the modified Bouc-Wen model is only considered for modelling MR fluid dampers.

3.12.1 Bingham Model

The Bingham model consists of a Coulomb friction element in parallel to a viscous damper as shown in Fig. 3.6 (Stanway et al. 1987, Spencer Jr et al. 1997).

Equation 3.23 gives the force generated by means of the MR damper:

$$f_{mr}^i = f_{mr}^y \, sgn\left(\dot{x}_{i+n} - \dot{x}_i\right) + c_0 \left(\dot{x}_{i+n} - \dot{x}_i\right) \tag{3.23}$$

where f_{mr}^y is the yield force, $\dot{x}_{i+n}$ and $\dot{x}_i$ are the velocities of adjacent buildings; subscript n denotes the total number of storeys of one of the adjacent buildings and c_0 is damping

Table 3.2 Design Parameters of the Large-scale MR Damper (Yang et al. 2002).

Stroke	± 80 mm
F_{MR}(**max**)/F_{MR}(**min**)	101@100 mm/sec
Cylinder bore (ID)	203.2 mm
Max. input power	< 50 W
Max. force (nominal)	200000 N
Effective axial pole length	84 mm
Coils	3 × 1050 turns
Fluid	2×10^{-10} sec/Pa
Apparent fluid	1.3 Pa sec
Fluid τ_0 max	62 kPa
Gap	2 mm
Active fluid volume	~ 90000 mm³
Wire	16 gauge
Inductance	~ 6.6 henries
Coil resistance (R)	3 × 7.3 ohms

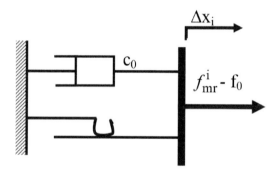

Fig. 3.6 Bingham model of controllable fluid damper (Stanway et al. 1987).

coefficient of the MR damper. In this model, there is no flow in the pre-yield condition because the material is rigid.

$$f_{mr}^y = f_{mra}^y + f_{mrb}^y u, \ c_0 = c_{0a} + c_{0b} u \qquad (3.24a–b)$$

The yield force and damping of the device in Eqs. 3.24a–b must be determined based on the applied voltage in order to provide a dynamic model having fluctuating magnetic fields (Jung et al. 2003).

3.12.2 Bouc-Wen Model

The Bouc-Wen model (Wen 1976) is numerically tractable and is used extensively for modelling hysteretic systems as shown in Fig. 3.7. The force created by the damper is shown as:

$$f^i_{mr} = \alpha z_{di} + c_0 (\dot{x}_{i+n} - \dot{x}_i) \tag{3.25}$$

where the evolutionary variable, z_{di} accounts for the history dependence of the response.

$$\dot{z}_{di} = -\gamma \left| \dot{x}_{n+i} - \dot{x}_i \right| z_{di} \left| z_{di} \right|^{n_d - 1} - \beta \left(\dot{x}_{n+i} - \dot{x}_i \right) \left| z_{di} \right|^{n_d} + A_c \left(\dot{x}_{n+i} - \dot{x}_i \right) \tag{3.26}$$

Some model parameters depend on the command voltage to the current driver. Hence, the relations between damping constants and voltage are proposed as follows:

$$\alpha = \alpha_a + \alpha_b u, \; c_0 = c_{0a} + c_{0b} u \tag{3.27}$$

The viscous damping parameters are linear manner to the applied voltage. Varying the constants in Eq. 3.26 can provide the smoothness of transition from the pre-yield to post-yield region to be controlled.

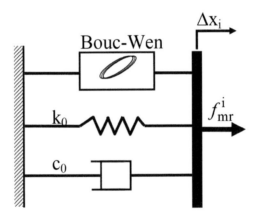

Fig. 3.7 Bouc-Wen model (Wen 1976).

3.12.3 Modified Bouc-Wen Model

The modified Bouc-Wen model as shown in Fig. 3.8 is used to simulate the dynamic behaviour of the MR damper that involves voltage-dependent parameters to model fluctuating magnetic fields.

$$f^i_{mr} = c_1 \dot{y}_i + k_1 (x_{i+n} - x_i - x_0) \tag{3.28}$$

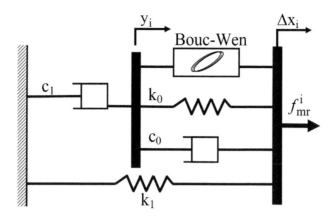

Fig. 3.8 Modified Bouc-Wen model for MR damper (Spencer Jr et al. 1997).

where the internal pseudo-displacement, y_i and the evolutionary variable, z_{di} are given by

$$\dot{y}_i = \frac{1}{(c_0 + c_1)} \left\{ \alpha z_{di} + c_0 \left(\dot{x}_{n+i} - \dot{x}_i \right) + k_0 \left(x_{n+i} - x_i - y_i \right) \right\}$$

$$\dot{z}_{di} = -\gamma \left| \dot{x}_{n+i} - \dot{x}_i - \dot{y}_i \right| z_{di} \left| z_{di} \right|^{n_d - 1} - \beta \left(\dot{x}_{n+i} - \dot{x}_i - \dot{y}_i \right) \left| z_{di} \right|^{n_d} + A_c \left(\dot{x}_{n+i} - \dot{x}_i - \dot{y}_i \right)$$

(3.29a–b)

where x_{n+i} and x_i are the displacement of the ith floor of Building B and Building A, respectively. Displacement of the MR damper Δx_i is computed, using relative displacement between two inline adjacent floors (i).

The x_0 is the initial displacement of spring of the accumulator stiffness k_1; k_0 is the stiffness at large velocities; c_0 and c_1 are viscous damping at large velocities and for force roll-off at low velocities, respectively; α is the evolutionary coefficient. Other shape parameters of the hysteresis loop are shown as γ, A_c, n_d and β in Eqs. 3.26, 3.27 and 3.29a–b. In this model, the following three model parameters depending on the command voltage u to the current driver are expressed as follows:

$$\alpha = \alpha_a + \alpha_b u, \ c_1 = c_{1a} + c_{1b} u, \ c_0 = c_{0a} + c_{0b} u$$

(3.30a–c)

Equation 3.31 is necessary to simulate the dynamics involved in reaching rheological equilibrium and driving the electromagnet in the MR damper. The dynamics are accounted for through the first-order filter

$$\dot{u} = -\eta (u - v_i)$$

(3.31)

where u is given as the output of a first-order filter which delays the dynamics of the current driver and of the fluid to reach rheological equilibrium; v_i is a command input voltage supplied to the damper at ith floor; f_{mr}^i is the damper force at ith floor level between the buildings. Parameter variables by optimal fitting of their model to test data are obtained by Spencer Jr et al. (1997). The optimised parameters for the three dynamic models that were determined to best fit the data derived from the experimental results of a 20-ton MR fluid damper in the study of Yang et al. (2002) are used herein. In order to find the data of a 100 ton (i.e. 1,000 kN) damper considered in this book, the experimental data of the 20 ton damper have been scaled up five times in the damper force and 2.5 times for the stroke of the device in a linear manner.

Chapter 4 deals with the active control system with actuators and semi-active strategies that are proposed to control the response of the structure after describing the dynamic model of the viscous and MR damper herein.

3.13 Summary

This chapter conducted the modelling of passive and active dampers for seismic response mitigation of a coupled building connected to each other by either passive or active dampers. Before describing the control strategies based on GA, non-dominated sorting genetic algorithm and pareto-optimal solutions in the following chapters, the modelling of MR dampers is investigated herein. One of the challenges to the application of MR dampers is use of an appropriate control algorithm to determine the command voltage of the MR damper. Many control algorithms are proposed to control the behaviour of MR dampers or other semi-active devices. Chapter 4 shows the control strategies for dampers that are proposed to control the response of the structure.

4

Algorithms for Designing
Optimal Control Force

4.1 Introduction

In order to design optimal control force, several optimisation methods based on the chosen objective function have been synthesised in this chapter. Typically, active and other control forces can be determined by using some control strategies, such as the Linear Quadratic Regulator (LQR), Linear Quadratic Gaussian (LQG), H_2, H_∞ norms, Fuzzy Logic Control (FLC). These control systems can be considered as objective functions for optimisation of passive and active control problems between adjacent buildings. This chapter includes a summary of the control strategies used in this book.

4.2 Linear Quadratic Regulator with Full State Feedback

This algorithm is one of the classic performance index used for active and other control devices of structures in the modern control theory. The optimal LQR method requires that all values of the state variables are available. Due to limitation in the number of sensors that could be installed in large structures for measuring the state variables, the use of this system is restricted for economical reasons. Firstly, an LQR algorithm with full state feedback is employed in this book. In order to construct the desired force in semi-active system, this research study is based on an LQR approach (Kirk 1970) which uses all states for feedback. For designing an LQR controller, the aim is to minimise the quadratic performance index subject to state Eqs. 3.9a–b without external excitation taken as the constraint (Levine and Athans 1970).

$$J = \frac{1}{2} \int_0^\infty \left[x^T \, Q \, x + F_{mr}^T \, R \, F_{mr} \right] d_t \tag{4.1}$$

Here, both $\mathbf{Q}$, the positive semi-definite state and $\mathbf{R}$, the positive define control input are weighting matrices in order to impose the importance of each term in Eq. 4.1. Hamiltonian can be formed by using the vector of regulated responses x in Eqs. 3.9a–b and control forces F_{mr}. By means of the help of the state and co-state vectors, Riccati differential equation in Eq. (4.2) can be obtained by the method discussed in Meirovitch (1992)

$$PA + A^T P + C_w^T \mathbf{Q} C_w - PB \, \mathbf{R}^{-1} B^T P = 0 \qquad (4.2)$$

where P is the solution of the algebraic Riccati equation. Optimal control force vector can be written as (Lewis and Syrmos 1995, Motra et al. 2011)

$$f_d = -B^T \, \mathbf{R}^{-1} PX = -\mathbf{K} X \qquad (4.3)$$

For multiple MR dampers, the control input is a vector, i.e. $f_d = [f_{d1} \, ... \, f_{di} \, f_{dm}]^T$ and $\mathbf{R} = [R]$; $\mathbf{K}$ is the full state feedback gain matrix for the deterministic regulator problem. As can be seen in Eq. 4.3, the resulting controller gain is time invariant. With all X measurements of the adjacent buildings, the control force is in the form of full state feedback controller.

4.3 Linear Quadratic Gaussian Regulator with Output Feedback

The H_2 optimal control theory is in the term of frequency domain interpretation of the cost function associated with time-domain state-space LQG control theory (Spencer et al. 1994, Dyke et al. 1996a). In this book, for semi-active control system, the H_2/LQG control algorithm is used through the reduced order model of adjacent buildings (Abdel Raheem et al. 2011). In many control algorithms, the aim is minimisation to a performance index based on the system variables with trading off regulation performance and control effort. The damper force in Eq. 4.3 can be found by minimising the performance index, subject to a second order system in Eqs. 4.4a–b.

Herein, the normal force of damper is used as the control input. The level of normal force required is determined by using an optimal controller based on an infinite horizon performance index form as:

$$J = \lim_{t \to \infty} \frac{1}{\tau} E \left[\int_0^\tau \left[\left(C_m X + D_m F_{mr} \right)^T \mathbf{Q} \left(C_m X + D_m F_{mr} \right) + F_{mr}^T \, \mathbf{R} F_{mr} \right] d_t \right]$$

or $\qquad\qquad\qquad\qquad\qquad\qquad\qquad\qquad\qquad\qquad\qquad\qquad$ (4.4a–b)

$$J = \lim_{\tau \to \infty} \frac{1}{\tau} E \left[\int_0^\tau \left[y_m^T \mathbf{Q} \, y_m + F_{mr}^T \, \mathbf{R} F_{mr} \right] d_t \right]$$

Both $\mathbf{Q}$ and $\mathbf{R}$ weighting matrices are for the vector of measured responses y_m in Eq. 3.9a–b and of control forces F_{mr}, respectively. Here, every element of the state vector is used in the feedback path ($C_w = C_m$). On the other hand, the number of sensors should be limited for economical reasons; the need of the output feedback, where not all states are available, is more pronounced (Arfiadi 2000). Many states in realistic systems are not easily measurable. The optimal controller in Eq. 4.3 is not implemental without the full state measurement (Levine and Athans 1970, Abdel Raheem et al. 2011). Hence, in this book a H_2/LQG controller is also employed as a nominal controller and the results are compared with the corresponding LQR controller. A state estimate can be formulated as $\hat{X}$ that $f_d = -\mathbf{K}\hat{X}$ remains optimal based on the measurements (Levine and Athans 1970, Abdel Raheem et al. 2011). Further, in the design of the H_2/LQG controller, the ground acceleration input, $\ddot{x}_g$ is taken to be stationary white noise. For design purposes, the measurement noise is statistically independent Gaussian white noise processes with $S_{\ddot{x}_g\ddot{x}_g}/S_{v_iv_i} = \gamma_g$ where $S_{\ddot{x}_g\ddot{x}_g}$ and $S_{v_iv_i}$ are the auto-spectral density function of ground acceleration and measurement noise, respectively. The nominal controller is represented as (Yoshida et al. 2003)

$$\dot{\hat{X}} = \left(A - L_g C_m\right)\hat{X} + L_g y_m + \left(B - L_g D_m\right)F_{mr}$$
$$L_g = \left(C_m S\right)^T$$
(4.5a–b)

where S is the solution of the algebraic Ricatti equation given in Eq. 4.6; $\hat{X}$ is the optimal estimate of the state space vector, X; L_g is the gain matrix for state estimator with the state observer technique, which is determined by solving an algebraic Riccati equation in control toolbox in MATLAB (R2011b).

$$SA^T + AS - SC_m^T C_m S + \gamma_g EE^T = 0$$
(4.6)

Based on selected displacement and velocity measurements, a Kalman filter is used to estimate the states. In order to produce an approximately desired control force f_d a force feedback loop is appended for inducing the MR device. A linear optimal controller $K_c(s)$ is designed to provide the desired control force f_d based on the measured responses y_m and the measured force F_{mr} as follows:

$$f_d = \mathbf{L}^{-1}\left\{-\mathbf{K}_c\left(s\right)\mathbf{L}\left(\begin{bmatrix} y_m \\ F_{mr} \end{bmatrix}\right)\right\}$$
(4.7)

where $L(.)$ is the Laplace transform. Although the controller $K_c(s)$ can be obtained from a variety of synthesis methods, the H_2/LQG strategies are conducted herein due to the stochastic nature of the earthquake ground motions and because of their successful application in other civil engineering control applications (Spencer et al. 1994, Dyke et al. 1996b, Abdel Raheem et al. 2011).

$$\mathbf{K}_c(\mathbf{s}) = \mathbf{K}\left[s\mathbf{I} - (A - \mathbf{LC})\right]^{-1}\hat{B}$$
(4.8)

65

where $\hat{B} = [\mathbf{L}\ \mathbf{B} - \mathbf{LD}]$. $\mathbf{K}$ is shown in Eq. 4.3 using the algebraic Ricatti equation given by Eq. 4.2 in the control toolbox in MATLAB (R2011b). The H_2 optimal control criteria defined in Section 4.4 can be numerically equivalent to the LQG optimal control criteria defined herein with appropriate selection of design weights in Eq. 4.4a–b (Lu 2001). Note that the damper is driven by the applied command input voltage (v). The states (x, y, z_d, u) are obtained via integration of Eqs. 3.9a–b, 3.29a–b and 3.31 using MATLAB module ode 45 based on 4th/5th order Runge-Kutta method. Then, the available damper force, $F_{mr} = [f_{mr}^1 \ \cdots \ f_{mr}^i \ f_{mr}^m]^T$ and desired force f_d are obtained via Eqs. 3.28 and 4.3, respectively. The schematic block diagram for implementations of LQR combined with a clipped voltage law (CVL) and H_2/LQG – CVL is illustrated in Fig. 4.1.

The conventional semi-active control strategy for an MR damper can be divided into two steps—primary and secondary controllers (Symans and Constantinou 1999). While the primary controller determines the best optimal force required for the MR damper, the input voltage of the MR damper is determined by secondary controller. The optimal force is clipped by the secondary controller in a consistent manner. This step is referred as the clipped optimal control strategy (Jansen and Dyke 2000) in Fig. 4.1.

Inverting the damper dynamics to obtain command voltage for a desired force is not possible from Eqs. 3.29a–b, 3.30a–c and 3.31. Hence, two methods—one LQR – CVL and the other based on H_2/LQG – CVL are used to obtain the voltage v based on desired and measured damper forces as described in Section 4.6.1.

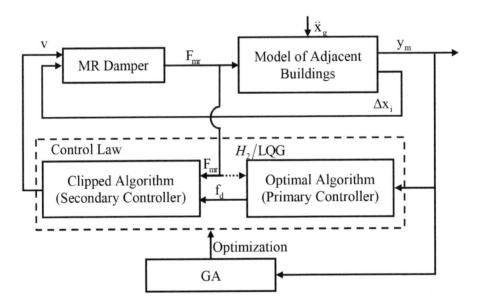

Fig. 4.1 Semi-active control block diagram of LQR – CVL and H_2/LQG – CVL.

4.4 H₂ Control

The use of H_2 optimisation procedure for the structural response under earthquake loading has been considered by many researchers in civil engineering application (Holland 1992, Spencer et al. 1994, Dyke et al. 1996b). The objective of H_2 methods is to minimise the transfer function of the closed loop system from external disturbances to a certain controlled output. The H_2 norm can be determined by

$$\left\|\hat{G}_{xw}\right\|_2 = \left(\frac{1}{2\pi} \int_{-\infty}^{\infty} tr\left(\hat{G}_{xw}(j\omega)\hat{G}_{xw}^*(j\omega)d\omega\right)\right)^{1/2} \tag{4.9}$$

where $\left\|\hat{G}_{xw}\right\|_2$ is H_2 norm transfer function from external disturbance $w = \ddot{X}_g$ to the controlled output x. While ω and j are frequency and imaginary, * and *tr* represent complex conjugate transpose and the trace, respectively. By Parseval's theorem, Eq. 4.9 in order to perform a physical interpretation of the H_2 norm is equal to the H_2 norm of impulse response

$$\left\|\hat{G}_{xw}(s)\right\|_2 = \left\|g_{xw}(t)\right\|_2 = \left(\int_{-\infty}^{\infty} tr\left(g_{xw}(\tau)g_{xw}^*(\tau)d\tau\right)\right)^{1/2} \tag{4.10}$$

where $g_{xw}(t)$ is the impulse response matrix given as

$$g_{xw}(t) = 0 \text{ when } t < 0$$
$$g_{xw}(t) = C_w e^{At}E \text{ when } t \geq 0 \tag{4.11a–b}$$

The regulated output *x* as the response to be kept small is chosen in relation to the performance index to be minimised.

$$x = C_w X \tag{4.12}$$

Substituting Eq. 4.11a–b into Eq. 4.10, the H_2 norm of impulse response can be obtained as given in detail in the study of Lu (2001).

$$\left\|\hat{G}_{xw}\right\|_2^2 = \int_0^{\infty} tr\left(C_w e^{At}EE^T e^{A^T t}C_w^T\right)d\tau \tag{4.13}$$

Rearranging Eq. 4.13 with defining $L_c = \int_0^t e^{A\tau}EE^T e^{A^T\tau}d\tau$ Leibniz's rule on the order of differentiation and integration, the controllability Gramians L_c can be obtained from the form of $\dot{L}_c = EE^T + AL_c + L_cA^T$ with $L_c(0) = 0$. Since A is asymptotically stable, the controllability Gramians L_c converges to a constant matrix. For infinite-horizon case $(t \to \infty)$, L_c is a constant matrix and $\dot{L}_c$ is zero. Because of the fact that the trace has the property *trace* (AB) = *trace* (BA) (see (Lu 2001)), the term $\int_0^{\infty} tr(Ee^{At}C_wC_w^T e^{A^T t}E^T)d\tau$ can also equal to $\left\| \hat{G}_{xw} \right\|_2^2$ (see Eq. (4.15a–b)). The H_2 norm transfer function from w to x can be computed, using (Lublin et al. 1996)

$$\left\| \hat{G}_{xw} \right\|_2 = \left[tr\left(C_w L_c C_w^T \right) \right]^{1/2} = \left[tr\left(E^T L_o E \right) \right]^{1/2} \tag{4.14}$$

where L_c and L_o which can be determined from the Lyapunov equations are the controllability and observability Gramians, respectively.

$$AL_c + L_c A^T + EE^T = 0 \tag{4.15a–b}$$
$$A^T L_o + L_o A + C_w^T C_w = 0$$

Note that displacements, velocities and absolute accelerations of adjacent buildings can be included to the controlled output defined in Eq. 4.12. By choosing the appropriate entry in the regulation matrix, certain regulated output that needs to be minimised can be imposed. For H_2 optimal feedback control,

$$\dot{X} = A_{cl}X + E\ddot{x}_g$$
$$U(t) = -Gy_m = -GC_m X = -G_z X$$
$$x = C_w X$$
$$J = \left[tr\left(C_w L_c C_w^T \right) \right]^{1/2} = \left[tr\left(E^T L_o E \right) \right]^{1/2} \Rightarrow Min. \tag{4.16a–g}$$
$$A_{cl}L_c + L_c A_{cl}^T + EE^T = 0 \quad \text{or}$$
$$A_{cl}^T L_o + L_o A_{cl} + C_w^T C_w = 0$$
$$A_{cl} = A - BGC_m = A - BG_z$$

Further, for feedback control system in active control systems, A matrix in Eqs. 4.15a–b will be replaced with A_{cl} in Eqs. 4.16a–g.

4.5 H$_\infty$ Control

To quantify the transfer functions, H_2 and H_∞ norms are usually used. In H_∞ controllers, the objective is to minimise the infinity norm of the transfer function from external disturbances to the regulated outputs. The H_∞ norm can be cast in the iterative manner. In this case, Hamiltonian matrix can be defined as:

$$\boldsymbol{H} = \begin{bmatrix} A + ER^{-1}D^T C_w & ER^{-1}E^T \\ -C_w^T \left(I + DR^{-1}D^T \right) C_w & -\left(A + ER^{-1}D^T C_w \right)^T \end{bmatrix} \tag{4.17}$$

where $R = \gamma^2 I - D^T D$. In this book, eigen values of this matrix in Eq. 4.17 are symmetric about the real and imaginary axes with $D = 0$; H_∞ norm can be computed in the following bisection algorithm:

(a) Select γ_u, γ_l so that $\gamma_l \leq \|\hat{G}_\infty\| \leq \gamma_u$

(b) If $(\gamma_u - \gamma_l)/\gamma_l \leq$ specified level (Tol.)

Yes Stop ($\|\hat{G}_\infty\| \approx \dfrac{1}{2}(\gamma_u + \gamma_l)$)

Otherwise, go to Step (c)

(c) Set $\gamma = (\gamma_u + \gamma_l)/2$ and test if $\|\hat{G}_\infty\| \leq \gamma$ using $\lambda_i(\boldsymbol{H})$

(d) If $\lambda_i(\boldsymbol{H})$ j$\mathbb{R}$, then set $\gamma_l = \gamma$, otherwise set $\gamma_u = \gamma$ and go to Step (b)

The resulting γ is the H_∞ norm to be determined. In the numerical solution, the computation of H_2 norm in Eq. 4.10 and H_∞ norm in bisection algorithm can be obtained by using lyap and norm commands in the MATLAB Control System Toolbox. In the graphical methods, the peak value is the singular value plot of the transfer function. The controller gains are then solved by using genetic based optimiser. In the clipped optimal controller, for each MR damper, the approach is to produce approximately a desired control force which is determined by means of H_2/linear quadratic Gaussian (LQG) and linear quadratic regulator (LQR) strategies. The next section describes the semi-active fuzzy control strategy for adjusting the MR dampers based on the input voltage of the MR damper after determining the ideal optimal force required for the MR damper.

4.6 Clipped Optimal Control

The clipped-optimal control method is used to solve an optimal control problem and to calculate the optimum force. Figure 4.2 illustrates the hysteretic behaviour of the MR damper model according to the input voltage. Non-linear force of damper is not directly controllable and applied voltage to the current driver can only be adjusted to reach the desired control force at each time step. Damping force produced by MR damper has also limited capacity and mostly cannot satisfy the calculated optimal control force.

Thus the applied voltage is set after computation of the optimal control force by a predefined control algorithm according to feedback data and measurement of damper force at each time in order to approach the MR damper control force to the desired optimal force. In other words, only the control voltage v_i can be directly controlled to adjust the force obtained by the device.

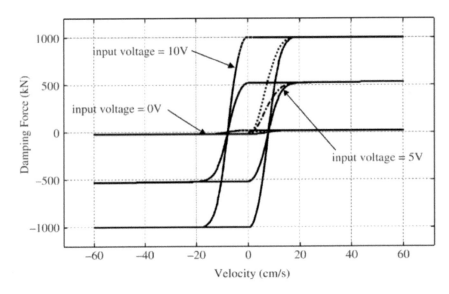

Fig. 4.2 Hysteretic behaviour of an MR damper.

4.6.1 Clipped Voltage Law

The input voltage, v, to the damper is obtained by using the CVL (Dyke et al. 1996b) and is described below. If these two forces are equal, then the applied voltage does not change. The applied voltage to the damper remains at present level. If the absolute of MR damper force is less than the absolute of the calculated optimal control force and both of them have the same sign, the applied voltage should be increased to its maximum value. Otherwise, the input voltage is set to zero. Clipped-optimal method can be summarised in the following equation:

$$v = V_{max} \mathbf{H}\{(f_d - F_{mr})F_{mr}\} \tag{4.18}$$

where V_{max} shows the maximum applied voltage that is associated with saturation of magnetic field in MR damper and $\mathbf{H}\{.\}$ is the Heaviside function. Voltage applied to the MR damper should be V_{max} when $\mathbf{H}\{.\}$ is greater than zero; otherwise, the command voltage is set to zero. Figure 4.3 indicates the schematic representation for implementation of control law block for CVL.

As a summary, in the clipped-optimal control algorithm, the command voltage varies to the value of either zero or the maximum value. In some cases, the effect of large changes in the forces applied to adjacent buildings to avoid high local acceleration values can be controlled by means of the time lag in the generation of the control voltage obtained by Eq. 3.31.

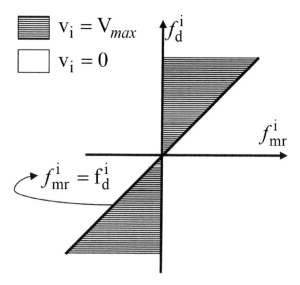

Fig. 4.3 Graphical representation of control voltage selection in the clipped optimal control algorithms.

Although the modified clipped-optimal control algorithm to reduce this effect is developed by Yoshida and Dyke (2004), the original clipped-optimal control algorithm is used in this book. Although significant studies have been made, based on active and semi-active control of building vibrations in seismic zones (Singh et al. 1997, Arfiadi and Hadi 2001, Yoshida et al. 2003, Aldemir 2010), the applications for intelligent controller, such as neural networks based control (Xu and Zhang 2002) and fuzzy logic control (Zhou et al. 2003), have not been addressed extensively. Algorithms for intelligent control have the advantage of not requiring a model of the system.

4.7 Fuzzy Logic Control Theory and Applications

Zadeh (1965) developed the fuzzy set theory in order to accommodate imprecision and uncertainty often presented in specific applications. Fuzzy logic control has been the focus of structural control engineers during the last two decades because of its sufficient inherent robustness to the closed loop system and ability to handle the non-linear dynamics of an MR damper (Battaini et al. 1998, Symans and Kelly 1999, Schurter and Roschke 2001). An optimal fuzzy logic controller to monitor the command voltage is also investigated in this book. The progress of the fuzzy controller is a more difficult and sophisticated procedure than that used in conventional control algorithms which are employed in this book, although fuzzy logic performs in the design of simple control algorithms (Kim and Kang 2012). For this reason, researchers are still faced with challenges when the fuzzy logic system is investigated for reduction of the structural vibrations. Fuzzy sets and rules require a good understanding of how to handle the non-linear dynamics of an MR damper.

Fuzzy set theory defines the way an input mapping to an output using verbose statements rather than mathematical equations for human reasoning (Nguyen et al. 1995), which involves terminology different from traditional mathematics (Symans and Kelly 1999). Because of the difficulty of establishing an accurate mathematical model, FLC can offer a simple framework for non-linear control laws. Using the clipped optimal, H_2/LQG and LQR control techniques, the change in command voltage sent to MR damper is not gradual or smooth. Furthermore, swift changes in voltage supply lead to a sudden rise in the external control force (Ok et al. 2007, Ali and Ramaswamy 2009, Kim and Kang 2012). Therefore, a fuzzy logic controller used in conjunction with MR damper is investigated to cover all voltage values in the range of zero and maximum MR damper voltages. The previous studies of FLCs mainly focused on various design parameters related to selection of membership functions and definition of the rule base to perform at the desired level (Yan and Zhou 2006, Pourzeynali et al. 2007). The membership functions and control rules of a fuzzy controller are usually determined by trial and error which is a tedious and time-consuming task. Because of the difficulty in determining the correlation between the structural responses and the command voltages, especially for a multi input multi output (MIMO) system, the design of the fuzzy controller requires optimally into GA. Figure 4.4a shows the proposed control strategy for integrated fuzzy logic and genetic algorithms.

As can be seen from Fig. 4.4b, the fuzzy logic has four modules: fuzzification, rule base, inference mechanism and defuzzification (Park et al. 2005, Ok et al. 2007). The detail of encoding fuzzy logic structure is given in the following chapter. As a summary, the basic

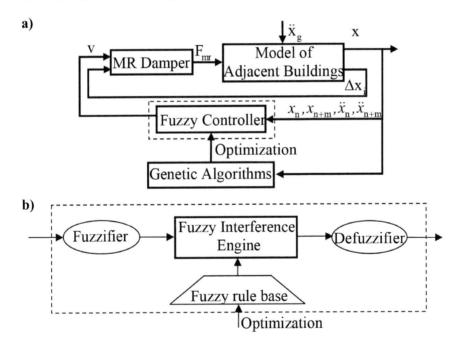

Fig. 4.4 Conceptual diagram of a) semi-active fuzzy logic control b) FLC components.

structure of a typical FLC is illustrated in Fig. 4.4b in which the components are defined as follows:

- Fuzzifier: takes the form of a crisp value which is then converted into fuzzified inputs using the input membership functions
- Fuzzy rules: constructed to achieve the control goal by specifying a set of if-premise-then-consequent statements
- Fuzzy interference engine: uses the fuzzy rules in the rule base module to infer fuzzified outputs from the fuzzified inputs
- Defuzzifier: operates on the fuzzified outputs obtained from the inference mechanism by converting into the required crisp control value

The most widely used fuzzy control inference R^i is the "if-then" rule, which can be written in Fig. 4.5 when two input data are used in their antecedent parts (Ahlawat and Ramaswamy 2003).

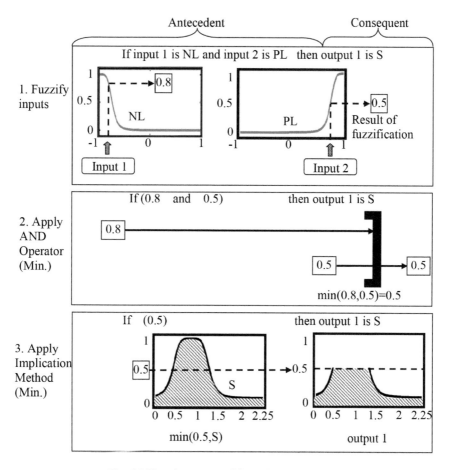

Fig. 4.5 First three steps of fuzzy logic controller.

An optimal design of fuzzy control rules and membership functions of the fuzzy controller is desired for efficiency. In this book, the shape and distribution of the defined membership functions are kept constant when the selection of fuzzy rules of the FLC system employs an adaptive method in GA. First, the input and output space of the system to be controlled are divided into fuzzy regions and the membership functions are defined for the design of an ordinary fuzzy controller. The integrated GA-FLC uses GA to derive proper rules from the initial rules (however, the initial rules are not necessarily needed in this book).

4.7.1 Integrated Fuzzy Logic Control Procedures

An adaptive method for selection of fuzzy rules of the fuzzy logic control system is conducted in GA in this section. Multi-input multi-output system can be complicated to combine for the adaptive method. A fuzzy correlation between the selected structural responses is optimally established, using GA. The foundation of this correlation is established for determining the corresponding command voltages for MR dampers according to the structural displacement responses of the highest floors for each building. Other parameters of the fuzzy controller, such as the shape and distribution of the membership functions are unchanged once defined. First, the input and output space of the system to be controlled is divided into fuzzy regions and the membership functions are defined for design of an ordinary fuzzy controller. The integrated GA-FLC architecture uses GA to derive proper rules from the initial rules (however, the initial rules are not necessarily needed in this study). Figure 4.6 shows graphically the conceptual diagram of the proposed fuzzy control strategy.

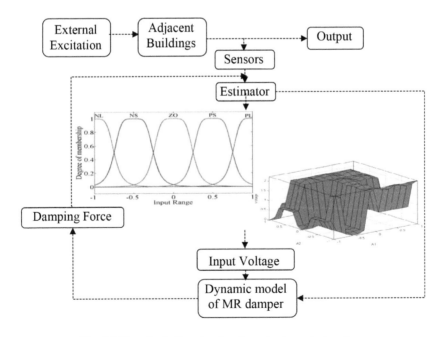

Fig. 4.6 Flowchart of semi-active fuzzy logic control system.

As can be seen from Fig. 4.7, the proposed fuzzy control strategy does not require a primary controller. The input voltage to the MR damper as input information from the response of the MR damper as output information can be obtained by the FLC. Composition of the fuzzy logic has four modules—fuzzification, rule-based, inference mechanism and defuzzification (Ok et al. 2007).

In order to be controlled into fuzzy regions and define the membership functions, the design of the fuzzy controller begins with selection of a range of input values. Here, a fuzzy controller has been defined, using a total of five membership functions for each of the inputs. A fuzzy set in this study is defined, using five abbreviations—NL (Negative Large), NS (Negative Small), ZO (Zero), PS (Positive Small) and PL (Positive Large). For the input membership functions, a reasonable range for each input value is selected. For example, the outermost membership functions can be rarely utilised if the range is too large. Conversely, if the range is too small, the innermost membership functions are rarely utilised. Utilising either the outermost or the innermost membership functions limits variability of the control system (Symans and Kelly 1999, Yan and Zhou 2006). In order to avoid this limitation, 70–80 per cent of the maximum uncontrolled displacement responses of the corresponding floors is taken as a reasonable range for each input in this study. The definition of the fuzzy output membership function abbreviations are as follows: ZO (Zero), S (Small), M (Medium) and L (Large) as shown in Fig. 4.8.

The crisp input is converted into fuzzified inputs using the input membership functions. After all the fuzzy rules are evaluated, the results of the rules are combined and defuzzified to a single number output. Figure 4.9 shows this process more clearly. The expansion and contraction of the horizontal axis in the input membership function directly influence to define whether the output of the fuzzy logic is large or small. The rule-base module is constructed by specifying a set of *if*-and-*then*-consequent statements. A fuzzy logic control system with two inputs and a single output represents each damper between adjacent buildings to be designed. This two-input-single-output case is chosen to simply clarify the

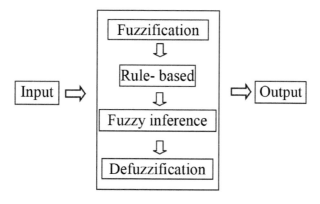

Fig. 4.7 Fuzzy control inference algorithm.

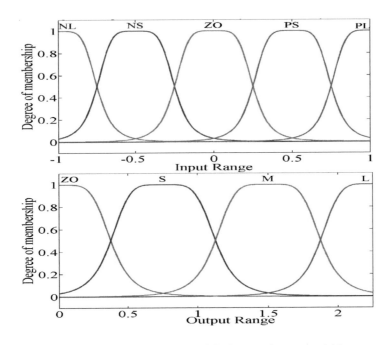

Fig. 4.8 Membership functions used for input and output variables.

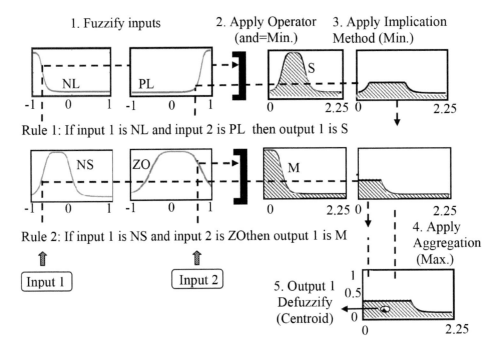

Fig. 4.9 Fuzzy logic controller in five steps.

basic ideas of how to represent a fuzzy rule base using bit-strings. Each input is divided into five fuzzy sets in this study.

A fuzzy rule base consists of 25 fuzzy rules for two fuzzy inputs. A five-by-five table with each cell to hold the corresponding outputs can be categorised for these rules. There are four choices in the voltage output corresponding to each rule. For example, using "IF-THEN" form, IF Input 1 is ZO and Input 2 is PL, THEN Output 1 is M, L, ZO or S. The *if-* part of the rule is called the antecedent which involves fuzzifying the input and applying any necessary fuzzy operators, while the ***then-*** part of the rule is called the consequent, known as implication. Each chromosome in the population consists of all the fuzzy rules and has the same input conditions but different output control signals assigned to bit-string. The bit-string is needed only to encode the output signals of the fuzzy rule.

There are totally 25 rules that form the fuzzy control rule base; thus 100 bits in a total can be used to represent a whole rule base for a single output. Each of the four consecutive bits are coded to represent the output for each rule. Each four bits from left to right represents the output linguistic variables—ZO, S, M and L, respectively. An output signal is selected by GA by setting the corresponding bit to 1 while the other three bits are 0s (Yan and Zhou 2006).

- IF Input 1 is NL and Input 2 is NL, THEN Output 1 is M
- IF Input 1 is NS and Input 2 is PS, THEN Output 1 is S
- IF Input 1 is ZO and Input 2 is ZO, THEN Output 1 is ZO
- IF Input 1 is PS and Input 2 is PL, THEN Output 1 is L
  (25 rules in total)

For example, the following rules can be represented by a chromosome as

$$0\ 0\ 1\ 0\ 0\ 1\ 0\ 0\ ...\ 1\ 0\ 0\ 0\ ...\ 0\ 0\ 0\ 1\ ...\ 0\ 1\ 0\ 0$$
(100 bits in total)

The current voltage obtained from the FLC combined GA is the same for all dampers installed between the buildings. The proposed integrated fuzzy logic and GA control strategy is especially suitable for designing an MIMO system. The multiple damper cases are similar to the single damper case, except that the bit number that represents the corresponding output voltages has a longer length. For example, while the same current voltages used for n_j number of MR dampers at the *j* storey level, different current voltages can be used for different storey levels. If different current voltages are needed throughout 10 storey levels, 250 rules are determined by GA using 1000 bits chromosomes.

However, the problem in this coding strategy is that it is hard to perform a mutation operation since it may result in more than one choice for each output in one rule. Therefore, in this study, only the selection and crossover operators are used to perform GA operations while neglecting the mutation operator because crossover is the main evolution operator in GA and mutation is secondary. Figure 4.10 shows one such input-output surface obtained in FLC optimisation at the beginning of simulation.

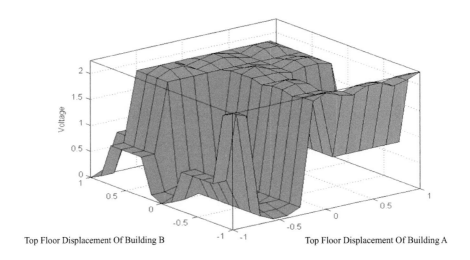

Fig. 4.10 FLC input/output surface.

Producing fuzzified outputs from fuzzified inputs in the inference mechanism can be provided by means of fuzzy rules in the rule base module. All fuzzy statements are resolved in the antecedent to a degree of membership between 0 and 2.25. After the entered and fuzzified input values, the fuzzy operator is applied to the antecedent and two fuzzified results are obtained as a single number. In this study, the *min* function for the *and* fuzzy operator (implication method) is used to determine the certainty of the fuzzy output variable for each fuzzy output. The truncated fuzzy set is defuzzified to assign a single value to the output command voltage by using the *max* function. For defuzzification, the centroid calculation is used in this study as the most popular defuzzification method which is the last component of the fuzzy logic. This module operates on the fuzzified outputs that are provided from the inference mechanism. A certain interface is required in order to relate the fuzzified values to the crisp output. In this numerical example, the input voltages are taken between the range of 0–2.25 V for fuzzy logic controller. Hence, a maximum value of 2.25 V is shown in the output membership function. This study adopts the centre of gravity (COG) method from among the defuzzification methods (Park et al. 2002). The COG method is defined as follows:

$$y_i = \frac{\sum\limits_{j=1}^{N_A} b_i^{(j)} \int \mu_{A_i}^{(j)}}{\sum\limits_{j=1}^{N_A} \int \mu_{A_i}^{(j)}} \tag{4.19}$$

where N_A is the number of rules activated from the inputs; $\mu_{A_i}^{(j)}$ is the value of the output membership function in the consequent statement for *j*th rule for *i*th input; $b_i^{(j)}$ is the centre of the output membership function $\mu_{A_i}^{(j)}$ and the integral of output membership function

$\int \mu_{Ai}^{(j)}$ represents the area of the output membership function. By means of the rule base in the inference mechanism, FLC determines the control voltage for the MR damper using the input functions.

4.8 Summary

Several control algorithms to design optimal control force and command voltage to MR dampers are discussed in this chapter. For seismic response mitigation, the objective of this book is to investigate the efficacy of the optimal additional damping and stiffness parameters due to instillation of the MR dampers using H_∞ optimisation. At the same time, the aim is to obtain optimum outputs (command voltages and number of dampers) of the controller for MR dampers by using LQR optimisation. For optimisation of semi-active control problems between adjacent buildings, several optimisation methods based on the chosen objective function are being synthesised in this book. The H_2/LQG norm is used to obtain the desired control force vector in the modern control theory, respectively. The clipped-optimal control method is to solve an optimal control problem and to calculate the optimum force. In the LQR control, the ground motion does not appear in the formulation as it is usually neglected. In H_2 and H_∞ control the disturbance is conducted as parameters in optimisation. They are represented in the modern state space approach even though H_2 and H_∞ controls are frequency domain controllers (Arfiadi 2000).

The fuzzy sets and rules that require a full understanding of the system dynamics must be correctly pre-determined for the system to function properly. Furthermore, in order to mitigate the responses of seismically subjected civil engineering structures, multiple MR dampers distributed between adjacent buildings should be used (Yan and Zhou 2006). Zhou et al. (2003) successfully applied an adaptive fuzzy control strategy for control of linear and non-linear structures. The authors found that the adaptive feature of a fuzzy controller has various advantages in the control of a building, including the MR damper system. To incorporate non-linearities in the model of the structure, the controller need not to be modified since the FLC can handle non-linearities.

In the next chapter, a framework on how to solve optimal static output feedback controllers using genetic algorithm is presented. For challenging the design of fuzzy control rules to control the MR damper voltage, a simple GA and Multi-Objective Genetic Algorithm (MOGA) are applied to the optimal design of FLC in this book. A combined application of genetic algorithms and Fuzzy Logic Controller (GFLC) are presented in the following chapter. In addition, the application of active and semi-active control concept to optimisation of passive control problems is also presented.

5

Genetic Algorithms for Single and Multi-Objective Optimisation Problems

5.1 Introduction

In general, engineers tend to design efficient systems, i.e. optimum designs. Modern structural design is a crucial step for the engineer to design in the optimum system. Safety is considered one of the most important design criteria, requiring the engineer to manage and design the members of a structure so that the whole structure will provide sufficient safety. Cost and feasibility are the other concepts in the design of structures. For optimisation of damper coefficients and number of dampers, genetic algorithm is briefly proposed in the following section. The idea behind the mechanics of Genetic Algorithm (GA) is that it simulates the Darwinian principle of 'survival of the fittest' and it is a probabilistic search method in nature. The GA uses binary coding to represent the design variable at its early development (Holland 1975, Goldberg 1989). This chapter contains the optimum design using genetic algorithm, NSGA-II and Pareto optimal solution, having discussed control algorithms. Basic mechanisms, components and advantages of genetic algorithms and multi objectives genetic algorithm as a structural optimisation method are defined in this section.

5.2 Basic Mechanics of Genetic Algorithm for Single Objective

The GA, which is inspired by natural selection and genetic evolution, is an adaptive heuristic optimisation technique for solving optimisation problems in engineering applications. The GA is one of the most flexible and effective methods to solve optimisation problems because of its robustness, constraint independence and parallel computation. In GAs, natural evaluation terms (the solutions to the optimisation problem) are represented as analogies (genetic strings) to real nature (chromosomes in nature). To understand the structure of GAs, a comparison between genetics terminology and the genetic algorithm terminology is given in Table 5.1. Binary string in GA can represent a candidate of a design variable. The GAs start off with a population of random candidates and advance towards better chromosomes by applying genetic operators in a specified coded string. After initialisation, the fitness of candidates is calculated to evaluate bit strings to each other according to the objective function. The genetic algorithm process is shown in Table 5.2.

The GA does not need the gradient information to optimise the cost function as the measure of the optimality is described by the fitness of individuals. This advantage provides the GA to be used for hard and complex optimisation problems with the capability of obtaining global optimum solution in a straightforward manner. The candidates undergo the selection process based on the fitness of each individual. In the selection process, the better chromosomes generate higher values than others and place them in the mating pool. The design variable (gene) of every individual (chromosome) in the population undergoes genetic evolution through crossover and mutation by a defined fitness function. In this chapter, the roulette wheel selection procedure maps the population in conjunction with the elitist strategy.

By using the elitist strategy, the best individual in each generation is ensured to pass to the next generation. After selection, crossover and mutation, a new population is generated in both coding GAs. This new population repeats the same process iteratively until a defined condition is reached. With the use of binary string (Goldberg 1989, Holland 1992, Arfiadi and Hadi 2011) or real number (Michalewicz 1996, Herrera et al. 1998), a candidate of a

Table 5.1 Comparison of Nature and Genetic Algorithm Terminology (Mitchell 1998).

Biological Name	Definition with GAs
Bit	Part of single variable
Chromosome	Encoding of a solution for objective function
Gene	Single variable encoding of a solution
Generation	An iteration of the genetic algorithm
Individuals	Feasible solutions
Population	Set of feasible solutions
Fitness of the individual	Fitness function for the quality of the solution
Evolution of population	Application of genetic operators
Initialisation	The progress by which a new generation of individuals is created randomly
Selection	The process for choosing which individual will reproduce
Crossover	A method of reproduction that combines the individual of multi parents to form a new individual
Mutation	When individuals are represented as bit strings, it means reserving a randomly chosen bit

Table 5.2 Structure of Genetic Algorithm.

Start (1)
Generation: $\tau \leftarrow 0$ % τ is iteration number
Initialise $G(\tau)$ % $G(\tau)$: Population for iteration
Evaluate $f(G(\tau))$ % $f(G(\tau))$: Fitness Function
while (not termination condition) do
start (2)
$\tau \leftarrow \tau + 1$
Perform operation of selection
Determine the number of crossover based on crossover rate p_c
Select the two parents $\widetilde{G}, \bar{G}$ from $G(\tau - 1)$
Perform crossover operation
Perform mutation operation for the whole population based on p_m
Insert a number of new random individuals replacing old individuals
Evaluate $f(G(\tau))$
end (2)
end (1)

design variable can be represented in GA. Not only does a GA provide an alternative method to solve problems, but it also produces results better than most of the other conventional methods. For a more thorough coverage of basic concepts of GAs, the reader is referred to Holland (1992), Goldberg (1989) and Arfiadi (2000). The details about the basic structure of genetic algorithms are briefly given as follows:

- Create a gene to find the potential solution for a particular problem
- Produce a method for an initial population of the potential solution
- Evaluate function in the role of the environment, evolving optimum solution by means of its 'fitness'
- Adjust the best individuals (genotypes) during reproduction, based on genetic operators
- Determinate parameters for probabilities by which genetic operators are applied

5.3 Binary Coding

A candidate of a design variable in GA can be represented either by using binary or real coding. In this section, binary coding is used for adjacent buildings connected by dampers. In the initial part of GA, using a string containing 0 and 1 represents individual as design variable. Briefly, the binary coding is presented here.

5.4 Chromosome Representation

By using a binary string having 1 or 0, the chromosome can be shown. The mechanics of GA start with creating an initial population of chromosomes as a set of candidates of initial design variables. The detail of the length of the individual is given in the thesis by Arfiadi (2000).

$$2^{(l_i-1)} < (U_i - L_i) \times 10^{P_i} \leq 2^{(l_i)} \tag{5.1}$$

The length of sub-chromosome (*n* bits) can be calculated based on upper (U_i) and a lower (L_i) bound values and the significant digit (P_i) of each design variable (*i*th). For example, Fig. 5.1 shows the total length of the chromosome, including the sum of the sub-chromosome length of each design variable.

It is clear that the length of a sub-chromosome depends on the required precision (P_i) of the design variable, as shown in Eq. 5.1. The l_i is the required length in bits used to represent the design variable. In binary coding, after initialisation, the conversion of binary strings into a real number of design variables is performed, using Eq. 5.2 (Hadi and Arfiadi 1998).

$$r_i = L_i + \frac{t_i \times (U_i - L_i)}{2^{nbits} - 1} \tag{5.2}$$

where r_i is the real number of a design variable; t_i is an integer mapping of a binary string which can be obtained by using

$$t_i = \sum_{j=0}^{l_m} h_j \times 2^j \tag{5.3}$$

in which the binary bit h_j is as follows: [h_r h_{r-1} ... h_1 h_0] and l_m is the length of sub-chromosome to represent a particular design variable.

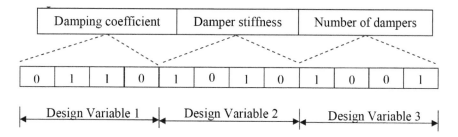

Fig. 5.1 An specimen chromosome with 3 design variables.

5.5 Selection Procedure

Chromosomes are selected by providing a higher probability of being selected for the mating pool to strings with a higher fitness value. For a single objective GA, the selection procedure used in this study is a roulette-wheel selection procedure in binary coding to select the individuals which pass for the next generation. Reproduction is processed in two stages. When the roulette wheel selection is used, the strings of the population are compared to the segments of a roulette wheel (Holland 1975). Each part of the wheel is sized proportionally to the fitness value of each individual as shown in Fig. 5.2. In the first stage, the fitness of each individual is evaluated and the sum of the fitness is calculated to determine the probability q_s of selecting each chromosome.

$$q_s = \frac{\text{fitness}}{\sum \text{fitness}} \tag{5.4}$$

The wheel is spun as many times as the total number of chromosomes in a population. In the second stage, the selection mechanism picks the highly-fitted chromosomes into the mating pool.

To perform this stage, the cumulative probabilities of selection of each individual are calculated (Hadi and Arfiadi 1998, Arfiadi and Hadi 2001, 2011). Each random number a_j ($j = 1, 2, ...$ popsize) between 0 and 1 is compared with the cumulative probability of selection of each chromosome q_j and when the random number is $a_j \leq q_j$, the jth individual will be selected.

$$q_j = \sum_{i=1}^{j} q_{s,i} \tag{5.5}$$

where $j = 1, 2, ...$ *popsize*. Here, popsize is the number of individuals in the population. The best individual is always selected for the next generation, using an elitist strategy by simply passing the best fitness individual on to the next generation. After the selection, the crossover and mutation operations may be performed. Fundamental aspects of the evolutionary search process of a genetic algorithm are the concepts of crossovers and mutations that modify the

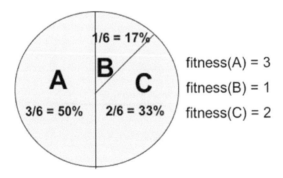

Fig. 5.2 Example of roulette wheel selection procedure for population having individuals.

84

chromosomes for the next generation of solutions. Simultaneously, these operators allow the exploration of new solutions while maintaining continuity of current solutions. For multi-objectives GA, the tournament selection is used to decide whether the chromosome in the population is to be selected or not in the mating pool.

5.6 Crossover

In this section, for reproducing new chromosomes, better features of solutions are transmitted to the optimisation problem. The crossover operator considers each individual in the mating pool taken by the selection operator to either perform upon the chromosome or not to recombine its genetic information into an offspring for the next generation. Some of crossover operators have been proposed and summarised into three types—one-point crossover operator, a two-point crossover operator and uniform crossover operator as shown in Fig. 5.3. For details of a simple crossover operation, refer to Holland (1975) and

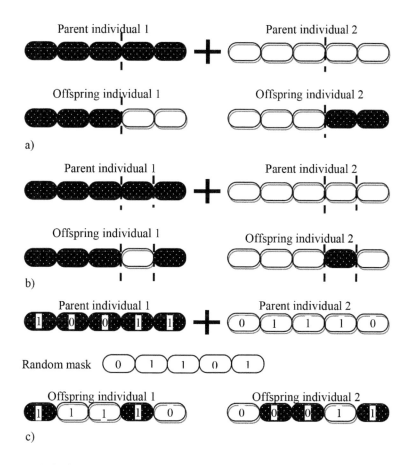

Fig. 5.3 Crossover a) single-point crossover operation b) two-point crossover, and c) uniform crossover operation.

Michalewicz (1996). A simple crossover is used as the main crossover operator for binary coding. Simple crossover randomly picks two parents from the mating pool and exchanges genetic information for one random split point in the chromosomes, as shown in Fig. 5.3a. If it happens at two points or *n* points, this operation is called two or *n* point crossover, as shown in Fig. 5.3b. Figure 5.3c shows a uniform crossover technique. In this operator, a random mask is randomly chosen with binary values (0 or 1). For each position in the offspring, the gene is copied from the other parent, according to a randomly generated crossover mask. Figure 5.3c shows how uniform crossover works. For example, parent individual 1 and individual 2 are 10011 and 01110, respectively, including random mask with 01101. Then, the children will be as 11110 and 00011. In this section, two chromosomes in Fig. 5.3a are chosen for a simple crossover operation (single-point crossover operation) if the random number is smaller than the crossover rate P_c.

Then, two new individual strings are formed by swapping the chain of strings at the end of the third gene as the crossover point. It may be completely different from their parents. Thus, the individual can be investigated for different solutions. With a simple crossover operator, swapping the chain of strings provides the best solution from good solutions (Holland 1992). After a simple crossover in the binary coding, mutation is performed in this chapter.

5.7 Mutation

In order to maintain diversity in the population, the mutation operator is used after applying the crossover operator. The binary bit at a certain position is flipped by means of this module, as shown in Fig. 5.4.

Extinct bits, which were at specific positions in the selection and crossover processes, can be brought back via mutation. It is the simplest genetic operator that can operate on all bits of strings of the tentative population. Generally, a small probability rate, P_m for mutation at each position is assigned. While the probability rate should be held at 1/p, where p is the number of decision variables for real coded GA (Deb and Agarwal 1995, Deb et al. 2002), according to Smith (1993), P_m should approach $1/L_c$, where L_c is the sum of sub-chromosome length of each string for binary coded GA, throughout the run (Smith 1993). Goldberg (1989) describes that lost points in the feasible search space may be regained for searching the search space. Thus, this genetic operator provides the algorithm from being stuck at local minima. Although the string will soon disappear when the mutated string has a low fitness, instead the mutation increases the individual's fitness. It may spread through the population and eventually lead to a new best fitness value. Mutation

Fig. 5.4 Mutation operation.

is a random operator whereby values of elements within a chromosome are modified. The $p_m \times$ (*nt* bits $\times$ popsize) bits will undergo a mutation operation if a random number n_r from the range (0–1) $< p_m$. Here, *nt* bits is the sum length of chromosomes (*n* bits). In this case, the mutation is run with the probability of mutation p_m. In order to maintain the variability of the population, mutation in binary coding is the random changing of 0,s to 1,s and vice versa.

5.8 Unknown Search Space

Figure 5.5 shows the flowchart of a Single Objective Genetic Algorithm (SOGA). For discovery of the larger unknown domain of the search space, the genetic operators are used for real-coded GA in this research study.

If the actual domain of the gain is unknown, this feature can be a benefit for optimal control problems. Multi-objective optimisation using evolutionary algorithms (MOEA) is popular and important from the point of practical problem solving. Although non-dominated sorting in genetic algorithms (Srinivas and Deb 1994) is very effective non-dominated based on genetic algorithm for multi-objective optimisation, it has been criticised because of its

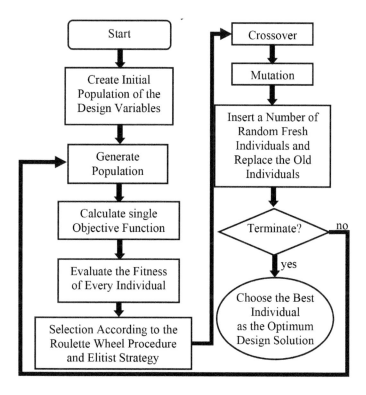

Fig. 5.5 The flowchart of single objective genetic algorithm.

computational complexity, lack of elitism and for choosing the optimal parameter value for sharing parameter.

5.9 Solution of Algorithms

A modified version, NSGA-II (Deb et al. 2002) was developed while considering elitism and no sharing parameter needs to be chosen, which has a better sorting algorithm. Once the population is initialised, the population is sorted out based on non-domination into each front. The first front being completely non-dominant is set in the current population and the second front is dominated by individuals in the first front only and the front keeps continuing. Two methods for multi-objective optimisation are investigated as non-dominated sorting and Pareto-optimal sorting. In both methods, the Pareto-optimal individuals from the previous generation are added before obtaining the Pareto-optimal individuals for the current generation.

5.10 Pareto-optimal Solutions

These algorithms attempt to search the possible design space for optimal design. The GAs begin with one or more populations containing a number of possible designs associated with corresponding fitness values. The algorithm reproduces new generations of solutions that generally have better fitness values than the solutions in the previous generations. Having better fitness values can be accomplished by performing operations of selection, reproduction, crossover and mutation of the coded design variables. The GAs in structural control applications initially and typically are used for a single structural response objective. Pareto-based Non-Dominated Sorting Genetic Algorithm (NSGA) is employed to get a solution of a multi-objective structural control problem. Multi-objective genetic algorithms provide a set of alternative solutions that trade different objectives against each other, generally known as Pareto-optimal solutions. An n-dimensional design variable vector $x = \{x_1, x_2, ..., x_n\}$ in the solution, space X can be given to define a multi-objective problem; $f(\bar{x}) = \{f_1(\bar{x}), ..., f_K(\bar{x})\}$ is given as a set of K objective functions to be minimised by a vector $\bar{x}$. The solution space X is generally restricted by a series of constraints, such as $g_j(\bar{x}) = b_j$ for $j = 1, ..., m$ and bounds on the decision variables. In order to define Pareto dominance and optimality for two-decision vectors x and y with objective vector f, the following mathematical expression can be defined for all the relative preference levels between solutions in the design space X:

$$x \succ y \text{ (x dominates y) if } f(x) < f(y)$$
$$x \succeq y \text{ (x weakly dominates y) if } f(x) \leq f(y) \tag{5.6}$$
$$x \cong y \text{ (x is indifferent to y) if } f(x)' \, f(y) \text{ and } f(y)' \, f(x)$$

In Goldberg's formulation (Goldberg 1989), the population of each generation is searched for non-dominated solutions. A solution x strongly dominates a solution y (x $\succ$ y), if solution x is strictly better than solution y in at least one objective. The solution x is no worse than a solution y (x $\succeq$ y) in all the K objectives. In order to obtain the true Pareto front, trading off among different objectives forms a set of solutions in multi-objective genetic algorithms. No weighting is specified by the user before or during the optimisation process. The optimisation algorithm provides a set of efficient candidate solutions from which the decision-maker chooses the solution to be used. Decision makers select the solution from the resulting Pareto-optimal set. The decision maker should finally decide on the relative importance of each objective function in order to get a single unique solution to be used as a solution for its original multi-disciplinary decision-making problem. Pareto optimality for minimisation of objective functions, f_1 and f_2 is illustrated in Fig. 5.6.

The set of all feasible non-dominated solutions in solution space is referred to as the Pareto optimal set, and for a given Pareto optimal set, the corresponding objective function values in the objective space are called the Pareto front. Solutions in the best-known Pareto set should be distributed uniformly and diversely over the Pareto front in order to provide the decision maker a true picture of the trade-offs. The ideal Pareto front should capture the whole spectrum of the Pareto front by investigating solutions at the extremes of the objective function space. An optimisation problem is considered for minimising two objective functions based on the vector of design variables. The Pareto front of non-dominated solutions on the Pareto curve are taken as optimal solutions since both the objective functions will increase as it leaves the surface. On the optimal surface, improvement in one objective function leads to degradation in the remaining objective function.

$$f_1(x_1) > f_1(x_2)$$
$$f_2(x_1) < f_2(x_2)$$

(5.7)

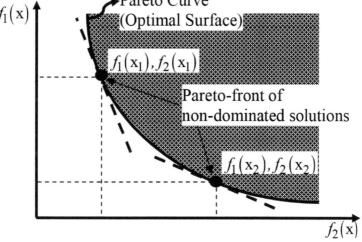

Fig. 5.6 Geometrical representation of the weight-sum approach in non-convex optimal curve case.

Any one of them can be considered as an acceptable solution since none of the solutions on the optimal surface is absolutely better than any other. Pareto-optimal or non-dominated solution set calls are such a set of optimal solutions. For Pareto-optimal solutions, a sharing technique is adopted to a non-dominated sorting procedure (Goldberg 1989). A non-dominated sorting genetic algorithm was presented in the study of Srinivas and Deb (1994).

5.11 Non-dominated Sorting Procedure

The initialised population is sorted out based on non-domination. The fast sort algorithm (Deb et al. 2002) is described as below:

- For each individual i in the main population $|P|$ (i = 1, 2,..., $|P|$) do the following
  - ➢ Initialise S_i. This set would contain all the individuals that are being dominated by i.
  - ➢ Initialise n_i = 0. This would be the number of individuals
  - ➢ For each individual, j in $|P|$ (j = 1, 2,..., $|P|$)
    - If i dominated j, then ($x_i \succcurlyeq x_j$ & **j ≠ i**)
      Add j to the set S_i, i.e. $S_i = S_i \cup \{j\}$
    - Else, if j dominated i, then ($x_j \succcurlyeq x_i$)
      Increase the domination counter for i, i.e. $n_i = n_i + 1$
  - ➢ If n_i = 0, i.e. no individual dominate i, then i belongs to the first front; set rank of individual i to one, i.e. i_{rank} = 1. Update the first front set by adding i to the front one, i.e. $P_1 = P_1 \cup \{i\}$
- This procedure is carried out for all the individuals in the main population $|P|$. Initialise the front counter to one k = 1, while the *k*th front is non-empty, i.e. $P_k \neq \emptyset$.
  - ➢ Initialise Q. This set would contain for storing the individuals for (k + 1)th front.
  - ➢ For each individual i ∈ P_k
    - For each individual j ∈ S_i
      $n_j = n_j - 1$ decrease the domination count for individual j
      If n_j = 0, then none of the individuals in the subsequent fronts would dominate j. Hence set j_{rank} = k + 1. Update the set Q with j, i.e. $Q = Q \cup \{j\}$
  - ➢ Increase the front counter by one k = k + 1
  - ➢ Now the set Q is the next front and hence P_k = Q.

This algorithm is better than the original NSGA (Srinivas and Deb 1994) since it utilises the information about the set that an individual dominates S_i and number of individuals that dominate the individual n_i. The crowding distance is assigned when the non-dominated sort is completed.

5.12 Elitism and Crowding Distance

In this chapter, NSGA-II, which was developed by Deb et al. (2002), is described as a popular and effective population-based heuristic search methodology for multi-objective optimisation. The operation of NSGA-II is shown in Fig. 5.7. A fast non-dominated sorting approach with elitism and diversity preserving mechanisms to enhance the performance of the evolutionary algorithm is used in this section. Firstly, a parent population P_t of random primal N chromosomes is formed. By using usual operators of GA, other N members are made. Then two populations are merged into one population of size 2N which is sorted, using a non-dominated sort algorithm (El-Alfy 2010, Fallahpour et al. 2012). The non-dominated sort generates a set of non-dominated fronts.

The solutions in the first non-dominated front are better than those in the second non-dominated front. After completing the non-dominated sort, non-dominated fronts are added sequentially to a new population of size N, starting with the best non-dominated front until the population is filled or reaches a non-dominated front that has more individuals than population. In the next section, another sort using a crowding distance metric is performed on this non-dominated front to select chromosomes which enhance the diversity of the solutions (El-Alfy 2010). Goldberg (1989) suggested a non-dominated sorting procedure in conjunction with a sharing technique. Consequently, Deb et al. (2002) presented the aim of crowding distance approaches to obtain a uniform spread of solutions along the best known Pareto front without using a fitness sharing parameter. For the multi-objective optimisation of MR dampers, this section adopts the NSGA-II (Deb et al. 2002) for a crowding distance method as follows (see Fig. 5.8).

- Step 1: Rank the population and identify non-dominated fronts. For each front, repeat Step 2 and Step 3

- Step 2: For each objective function, sort the solutions in the related front in an ascending order. Let x represents the *i*th solution in the sorted list with respect to the objective function. Assign

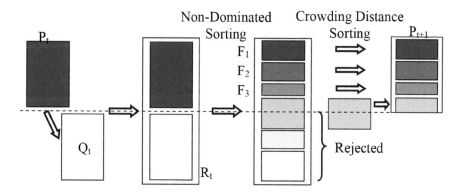

Fig. 5.7 Illustration of the operation of NSGA-II.

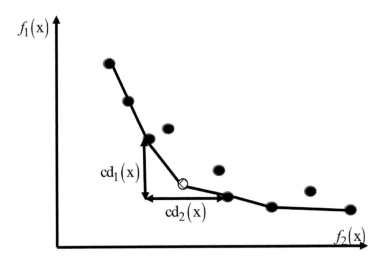

Fig. 5.8 Diversity methods used in multi-objective GA.

$$cd_k(x_{[i,k]}) = \frac{f_k\left(x_{[i+1,k]}\right) - f_k\left(x_{[i-1,k]}\right)}{f_k^{max} - f_k^{min}} \tag{5.8}$$

- Step 3: In order to obtain the total crowding distance $cd_k(x)$ of a solution x, sum the solution crowding distances with respect to each objective in Eq. 5.9

$$Cd_k(x) = \sum_k cd_k(x) \tag{5.9}$$

where f_k is a goal function and f_k^{max} and f_k^{min} are maximum and minimum for this function as shown in Fig. 5.8. The main advantage of the crowding method described above is that a measure of population density around a solution is computed without requiring a user-defined parameter.

In NSGA-II, this crowding distance measure is used as a tie-breaker as in the selection phase. Two solutions x and y are randomly selected. If the solutions are in the same non-dominated front, the solution with the higher crowding distance wins. Otherwise, the solution with the lowest rank is selected.

The selection is carried out by means of a crowded-comparison-operator using a relation $\prec_n$ as follows:

- The rank of i is better (less) than the rank of j, i.e. i and j belong to two different non-dominated fronts
- The ranks of i and j are same and i has higher crowding distance than j solution. This means that if two solutions belong to the same non-dominated front, the solution situated in the less crowded region is selected

The chromosomes are selected using a binary tournament selection with crowded comparison operator.

5.13 Simulated Binary Crossover

Simulated binary crossover which was proposed by Deb and Agarwal (1995) is given below:

- Initialise the children to be null vector with related to the probability perform crossover $p_c = 0.9$. Select two different parents (≥ 1) and get the chromosome information for each randomly selected parent
- The offspring values $C_{1,k}$, $C_{2,k}$ with kth component are calculated, based on the selected parent $P_{1,k}$, $P_{2,k}$ with a random number generated in Eq. 5.11a–b.

$$C_{1,k}(x) = \frac{1}{2}\left[(1-\beta_k)\times p_{1,k} + (1+\beta_k)\times p_{2,k}\right]$$

$$C_{2,k}(x) = \frac{1}{2}\left[(1+\beta_k)\times p_{1,k} + (1-\beta_k)\times p_{2,k}\right]$$

(5.10a–b)

$$\beta_k = (2u_j)^{1/(mu+1)} \quad \text{when } u_j \leq 0.5$$

$$\beta_k = \left(\frac{1}{2\times(1-u_j)}\right)^{1/(mu+1)} \quad \text{when } u_j > 0.5$$

(5.11a–b)

where β_k is a distribution factor with kth component; mu is distribution index for crossover as 20 (Deb and Agarwal 1995, Deb et al. 2002)

- Evaluate the generated element that is within the specified decision space, or else, set the offspring values to the appropriate upper or lower limit. Then evaluate the objective function of the offspring.

5.14 Polynomial Mutation

For real coded NSGA-II, mutation operator is based on polynomial mutation with a mutation rate of $p_m = 0.1$. After selecting the random parent, the individual information is taken for the selected parent:

$$C_k = c_k + \delta_k$$

(5.12)

where c_k is the parent and C_k is the child with kth component. The δ_k is small variation calculated from a polynomial distribution by using

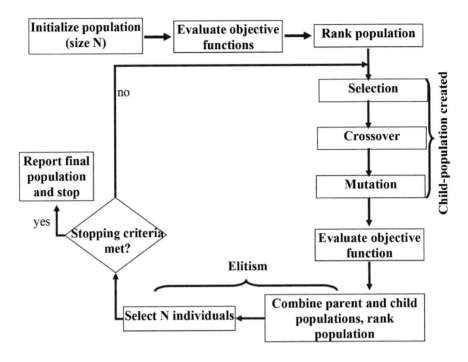

Fig. 5.9 Flowchart of NSGA-II.

$$\delta_k = (2r_k)^{1/(\eta_m+1)} - 1 \quad \text{when } r_k < 0.5$$

$$\delta_k = 1 - (2(1-r_k))^{1/(\eta_m+1)} \quad \text{when } r_k \geq 0.5$$

(5.13a–b)

As done in Section 5.13, the generated element is evaluated to be within the decision space. As shown in Fig. 5.9, this section follows the flowcharts of NSGA-II developed by Deb et al. (2002).

5.15 Evaluation of Genetic Algorithms

There are four key differences between Genetic Algorithms and other traditional search methods. As Goldberg (1989) summarised, the first one is that with the help of encoding of a feasible set, GA can work rather than the solution set itself. Secondly, GAs converge from multiple points on to a solution rather than from a single point. As the GA uses a population of points to carry its search and evaluates the entire population in each generation, it provides an opportunity to generate a set of non-dominated designs in one run (Ahlawat and Ramaswamy 2002, Ahlawat and Ramaswamy 2003). The third is that GAs work with a fitness function that is not based on any assumptions knowledge of the search space. Finally, GAs work with the help of probabilistic rules.

5.16 Multi Objective Genetic Algorithm Procedures

Figure 5.10 shows the optimisation tool for Pareto solutions in MATLAB. The NSGA-II approach (Deb et al. 2002) additionally requires the 'rank-based sorting' after the fitness assignment progress. It ranks the non-domination of individuals in the population based on the vectors of multi-objective functions. Then the set of solutions corresponding to the most non-dominated set is saved during the elitism operation. In the present form of the proposed framework, the control procedure uses the discrete input variables to compute a discrete output.

Here, the multi-objective optimisation problem for the MOGA case is the minimisation of three objective functions: (1) the number or dampers (*ndi*), (2) expected voltage of damper (*vdi*) and (3) a measure of the normalised maximum floor displacement relative to the ground or the peak drift. Specifically, the MOGA optimisation problem is defines as follows:

$$
\text{Minimise} \quad
\begin{cases}
J_1 = \text{ndi} \\
J_2 = \text{vdi} \\
J_3 = max\left\langle \dfrac{\left|x_i(t)\right|}{x^{max}} \right\rangle or \quad max\left\langle \dfrac{\left|d_i(t)\right|}{d^{max}} \right\rangle
\end{cases}
\tag{5.14}
$$

These objective functions are to use the MOGA to simultaneously minimise. A 'niche-based sorting' procedure is performed to provide a diverse search direction along the Pareto-optimal surface by degrading the fitness of densely populated individuals. Based

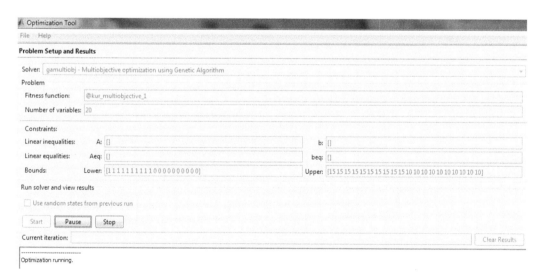

Fig. 5.10 Optimisation tool for pareto solutions in MATLAB.

on the fitness values, the next generation of the population is produced through selection, crossover and mutation operations.

5.17 MATLAB Program

The Fuzzy Logic Toolbox is an additional software for use with SIMULINK and MATLAB which permit a fuzzy logic controller to be placed within a SIMULINK model. Numerical examples for adjacent buildings in Chapter 6 are performed on i7-2630QM @2.9 GHz computer running MATLAB R2011b. The GA built on the MATLAB numeric computing environment is integrated into the SIMULINK block to simulate controllers.

5.18 Summary

The optimisation of structural control problems, using SOGA and NSGA-II in MOGA, are discussed in this chapter, where the binary as well as real coding GA are presented. The simple GA procedure is also slightly modified here. Some fresh individuals after selection, crossover and mutation are inserted into the population and inserting new individuals can help in exploring new candidates of the design points. Although a bad fitness individual can pass into the next generation with inserting new individuals, the average fitness of current population will be better than the average fitness of the previous population. As the optimum design, the best design points can be obtained in the final generation by copying the highest fitness value into the next generation. Chapter 6 shows the optimum design examples.

6

Verification of the Approximate and Rigorous Models for Adjacent Buildings of Different Dimensions

6.1 Introduction

A four-storey two-asymmetric building used by Jui-Liang et al. (2009) is selected for the SSI system, which is a modification of the examples adopted by Thambirajah et al. (1982, 1983). The properties of the SSI system are briefly defined in this chapter.

6.2 Model Description

The rectangular building having dimensions of 12 m × 15 m with the larger side being parallel to the y axis is used. The rates of e/r and f/r are 0.3 for each storey and the total height of the building from the base to the top floor is 30 m. The ratio of mass of base to the mass of any floor is taken as 3.0. The masses and translational stiffness (in the x axis or N-S direction) of floors are listed in Table 6.1. Each floor has a loading intensity of 1.4 kN/m^2.

The values of the translational stiffness in the y axis and torsional stiffness are given by:

$$\frac{k_{yj}}{k_{xj}} = \beta_y , \quad \frac{k_{\theta j}}{r^2 k_{xj}} = \beta_t \tag{6.1}$$

where β_y and β_t are constant values and the values of ratio are taken as 2.0 and 1.8, respectively (Jui-Liang et al. 2009). The damping ratio of the fundamental mode is 2 per cent herein.

Table 6.1 Mass and Translational Stiffness of the Four-Storey Building (Jui-Liang et al. 2009).

Properties	1st Floor	2nd Floor	3rd Floor	4th Floor
m_j (kg)	24465	24465	24465	24465
$k_{xj} \times 10^6$ kN/m	0.262	0.255	0.240	0.219

6.3 Selected Soil-Structure System

The properties of SSI system for the asymmetric building used by Thambirajah et al. (1983) and Jui-Liang et al. (2009) is the same with the properties of SSI system for the couple buildings as shown in Section 10.3.2. Two different SSI systems are used for the building with resting on the soft (Case I) and hard (Case IV) soils. Both systems were subjected to the 1940 El Centro earthquake. The impedance values of the SSI system for both the soft soil and hard soil are given in Table 6.2.

As can be seen from Table 6.2, K_ψ and C_ψ are the same as K_ϕ and C_ϕ respectively. The radius of base mass, r_o, for the impedance values of SSI system calculated by Jui-Liang et al. (2009) was taken as the radius of a circle having the same area of the building dimension. The impedance values of the coupled buildings for the SSI system modelled in Section 10.3.2 are determined with the radius of base mass for 5 DOF based on the calculations used by Richart et al. (1970) as given in this section. The proposed approximation method for modal response history analyses using MDOF modal equations and the direct integration method for the equation of motion for the whole SSI system are solved for the building by means of the ordinary differential equation solver of MATLAB. Furthermore, the findings of both the methods obtained by Jui-Liang et al. (2009) are compared with results obtained by current research study of this book.

$$M_a = \begin{bmatrix} m_1 & & & & \\ & m_2 & & & \\ & & m_i & & \\ & & & m_N \end{bmatrix}; \quad M_b = \begin{bmatrix} m_1 & & & & \\ & m_2 & & & \\ & & m_j & & \\ & & & m_s \end{bmatrix} \tag{6.2}$$

in which m_i is the mass of the first floor of Building A; m_j is the mass of the jth storey of Building B. Storey Stiffness and Eccentricities are given below.

$$k_{xi}^a = \sum_{j=1} k_{xji}^a, k_{yi}^a = \sum_{j=1} k_{yji}^a \tag{6.3}$$

$$k_{xj}^b = \sum_{i=1} k_{xij}^b, k_{yj}^b = \sum_{i=1} k_{yij}^b \tag{6.4}$$

Table 6.2 Impedance Properties for Cases I and IV (Jui-Liang et al. 2009).

Building types	K_T(kN/m)	K_θ(kN m)	$K_{\psi,\phi}$(kN m)	C_T(kN s/ m)	C_θ(kN m s)	$C_{\psi,\phi}$(kN m s)
Case I	3.03×10^5	1.88×10^7	1.41×10^7	2.03×10^4	2.52×10^5	4.61×10^5
Case IV	6.45×10^6	4.00×10^8	3.00×10^8	9.37×10^4	1.16×10^6	2.13×10^6

where k^a_{xji} and k^a_{yji} are the lateral stiffness of the jth resting element of the ith storey in the longitudinal and transverse directions, respectively; k^a_{xi}, k^a_{yi}, k^b_{xj} and k^b_{yj} are the total lateral stiffness of the ith and jth storey, respectively.

$$k^a_{\theta Mi} = \sum_{j=1} k^a_{xji} y^2_j + \sum_{j=1} k^a_{yji} x^2_j \qquad (6.5)$$

$$k^b_{\theta Mj} = \sum_{i=1} k^b_{xij} y^2_i + \sum_{i=1} k^b_{yij} x^2_i \qquad (6.6)$$

where x_i, y_i, x_j and y_j are the distances of the ith resting element from the centre of the mass in the x and y directions for Building B and of the jth resting element from the centre of the mass in the x and y directions for Building A respectively. Equations 6.5 and 6.6 show the torsional stiffness of the related storey in the buildings. The static eccentricities in the x and y directions are illustrated as:

$$e^a = \frac{\sum_j x_j k^a_{yji}}{k^a_{yi}}, \; f^a = \frac{\sum_j y_j k^a_{xji}}{k^a_{xi}} \qquad (6.7)$$

6.4 Validation of Analytical Results of the Approximate Method

Another four-storey asymmetric superstructure resting on a rigid base modelled as the reference building (RB). As the benchmark, the reference building will indicate how the SSI affects the dynamic properties of the building. In order to investigate the validation of current study, first six periods of modal vibration are compared with those periods obtained by Jui-Liang et al. (2009) as shown in Table 6.3.

As can be clearly seen from Table 6.3, the first six periods obtained by Jui-Liang et al. (2009) to the periods in the current study are very close to each other for both Case I and Case IV. It appears from Table 6.3 that the modal vibration periods of Case IV are almost unaffected when compared with Case I. While the ratios of Case I to the reference building associated with modal vibration periods range from 1.455 to 1.963, the range from 1.00 to 1.045 is for the ratios of Case IV of the reference building. Table 6.4 shows that the first six natural frequencies of the 8-storey building-foundation system used by Thambirajah et al. (1982) are also same as compared with the current study. As a validation of the current study, Fig. 6.1(a–c) and Fig. 6.2(a–c) show the mode shapes of the reference building,

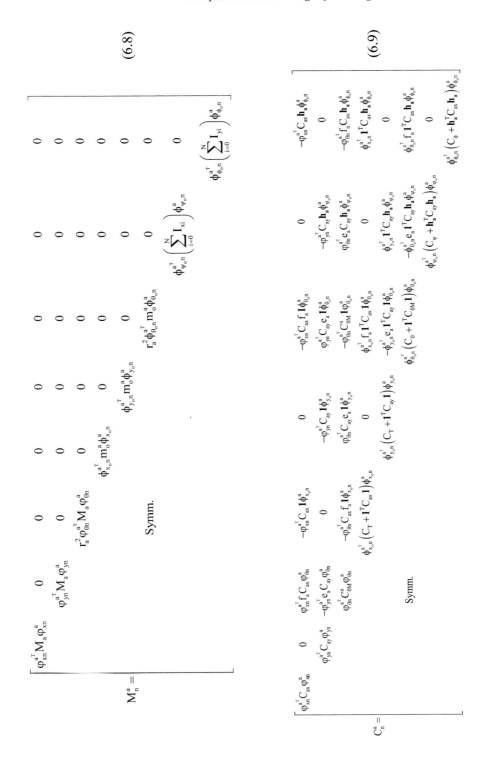

(6.8)

(6.9)

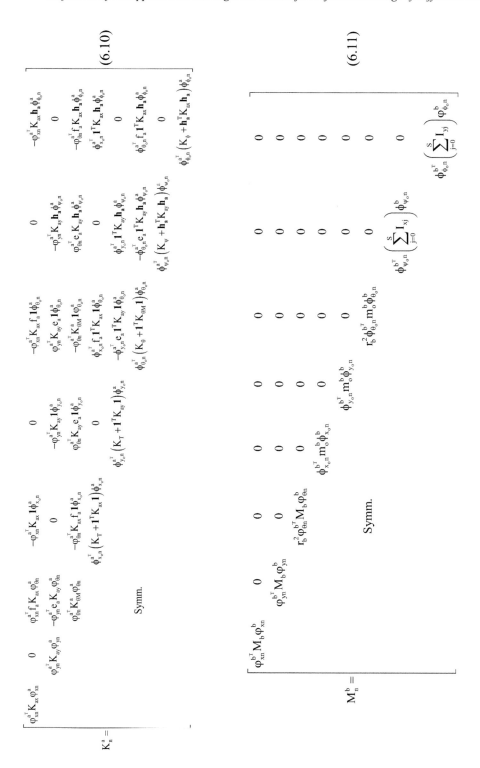

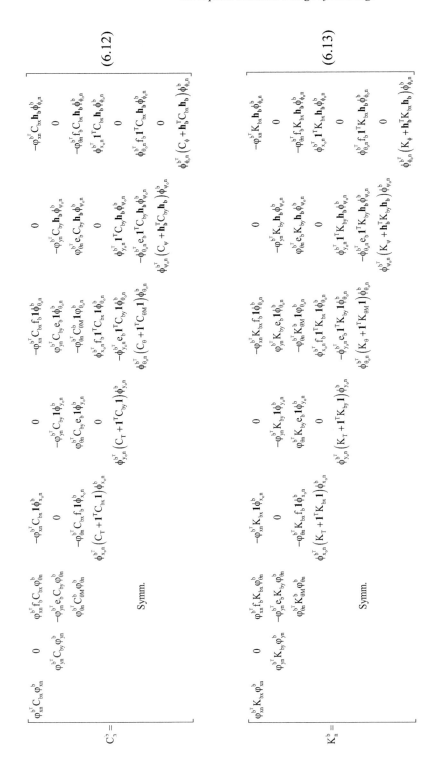

Table 6.3 First Six Periods of Modal Vibration Obtained by Jui-Liang et al. (2009) and Current Study.

Cases / Periods	Case I	Case I*	Case IV	Case IV*	RB	RB*	Case I* / RB*	Case II* / RB*
T_1 (sec)	0.281	0.281	0.196	0.196	0.192	0.192	1.464	1.021
T_2 (sec)	0.253	0.255	0.154	0.154	0.150	0.150	1.700	1.027
T_3 (sec)	0.161	0.163	0.117	0.117	0.112	0.112	1.455	1.045
T_4 (sec)	0.107	0.107	0.069	0.069	0.069	0.069	1.551	1.000
T_5 (sec)	0.104	0.106	0.054	0.054	0.054	0.054	1.963	1.000
T_6 (sec)	0.069	0.070	0.045	0.045	0.045	0.045	1.533	1.000

* Obtained by current study; RB: reference building

Table 6.4 First Six Natural Frequencies of 8-Storey Building-Foundation System Used by Thambirajah et al. (1982) and the Current Study.

Cases / Frequency	Case I	Case I*	Case IV	Case IV*	RB	RB*	Case I* / RB*	Case II* / RB*
W_1 (Hz)	0.698	0.685	0.788	0.787	0.794	0.792	0.865	0.994
W_2 (Hz)	0.783	0.787	0.932	0.931	0.941	0.940	0.837	0.990
W_3 (Hz)	1.120	1.107	1.224	1.222	1.233	1.231	0.899	0.993
W_4 (Hz)	1.896	1.874	1.942	1.937	1.943	1.941	0.965	0.998
W_5 (Hz)	2.227	2.175	2.301	2.297	2.306	2.303	0.944	0.997
W_6 (Hz)	2.905	2.795	3.013	3.007	3.019	3.015	0.927	0.997

* Obtained by current study; RB: reference building

while Fig. 6.1(d–f) and Fig. 6.2(d–f) indicate the mode shapes of Case I obtained via the approximate method by both current study and Jui-Liang et al. (2009), respectively. It can be noted that the radius of the first six natural frequencies of Case I with those of the reference building is smaller than the radius of frequencies of Case IV to the frequencies of the reference building in Table 6.4.

The periods of the building resting on the hard soil are significantly close to the reference building resting on a rigid base. Referring to Fig. 6.1, the offsets and slopes of the thin lines representing the translations and rotations of the base are the same as those of the lines in Fig. 6.2, respectively. The first to the third mode shapes of the reference building and Case I (System I) obtained by Jui-Liang et al. (2009) in Fig. 6.2 are in agreement with those obtained via the approximate method in the current study in Fig. 6.1. Both Fig. 6.1 and Fig. 6.2 show that the first three mode shapes of dominant motion are *x*-translation, *y*-translation and rocking in *y*-direction (Balendra et al. 1982, Jui-Liang et al. 2009).

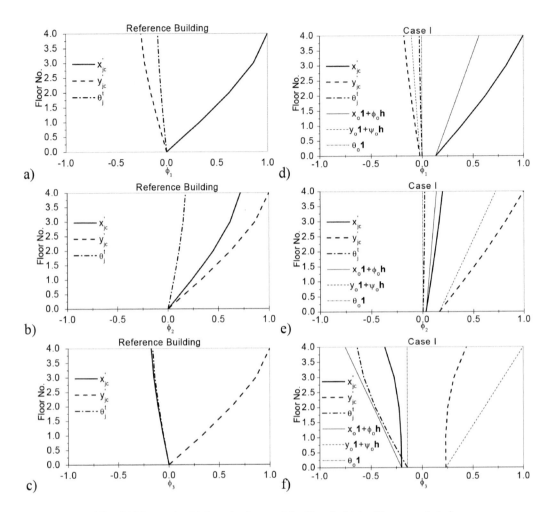

Fig. 6.1 First to the third mode shapes of the Case I obtained by current study.

The modal response histories in the first three modes of the building subjected to the El Centro 1940 earthquake are represented in Fig. 6.3 in order to compare the findings obtained by Jui-Liang et al. (2009) in Fig. 6.4. The modal displacement-time relationships of each eight degrees of freedom of Case I for each mode are clearly different because of the important soil structure effects in both Fig. 6.3(a–c) and Fig. 6.4(a–c).

On the other hand, it can be observed from Fig. 6.3(d–f) and Fig. 6.4(d–f) that the modal responses of Case IV for the eight DOFs are similar in manner. It implies that the modal displacement histories using the equivalent SDOF modal equations of motion represented by a SDOF modal system resting on a rigid base (see Fig. 2.5) can be same as the results of

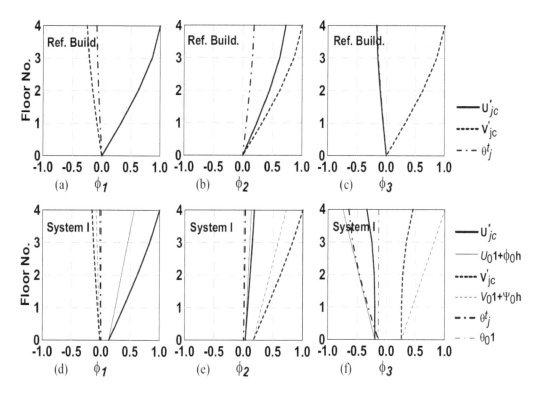

Fig. 6.2 First to the third mode shapes of the reference building and Case I (System I) obtained by Jui-Liang et al. (2009).

response histories obtained by the MDOF modal equations of motion (Balendra et al. 1982, 1983, Balendra and Koh 1991).

Furthermore, the response histories of the elements of modal coordinate in Fig. 6.3(c) are slightly smaller than the response of those elements referred to in Fig. 6.4(c). The modal responses of Case I for the third mode in both Fig. 6.3(c) and Fig. 6.4(c) are mostly affected due to the SSI effects in conjunction with the phenomena of out-of-phase vibrations among the eight DOFs in each vibration mode (Jui-Liang and Keh-Chyuan 2007, Jui-Liang et al. 2009).

6.5 Validation of Analytical Results of the Rigorous Method

Figure 6.5 and Fig. 6.6 show the total response histories of Case I and Case IV for both the approximate method (App.) using the MDOF modal equations of motion in Eq. 2.36.

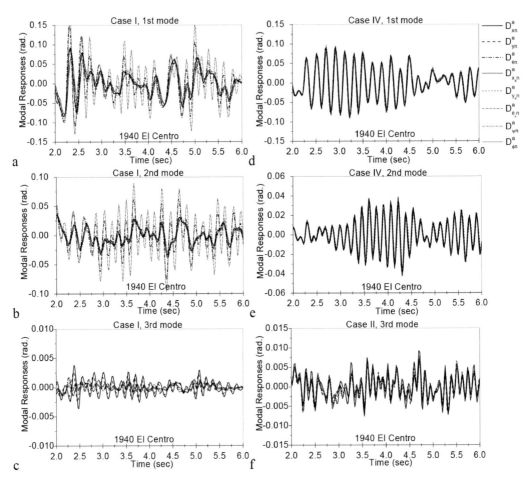

Fig. 6.3 Modal response time histories of the first to third modes for both Case I and Case IV under the excitation of the 1940 El-Centro earthquake in current study.

The rigorous method (Rig.) using the equation of motion for the whole SSI system in Eq. 2.22 is illustrated in Fig. 6.5, compared with those obtained by Jui-Liang et al. (2009). It should be noted that only the first three vibration modes are considered for the approximate method as shown in Eq. 2.40a–b.

Except Fig. 6.5(d–e), not only peaks but also the phase of the total response histories of Case I and Case IV obtained by the current study for both the methods are in agreement with those obtained by Jui-Liang et al. (2009). However, in Fig. 6.5(d–e), the peaks of total

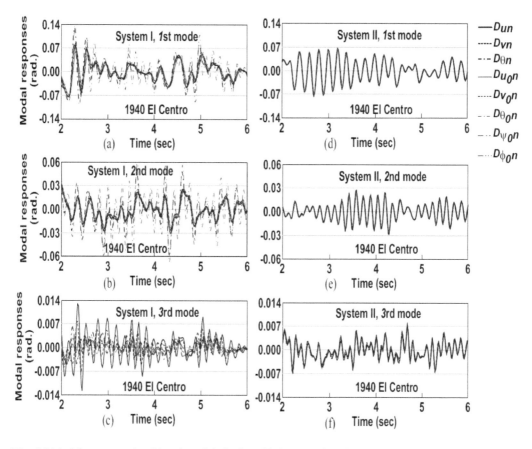

Fig. 6.4 Modal response time histories of a) the first; b) the second; and c) the third mode for System I (Case I); d) the first; e) the second; and f) the third mode for System II (Case IV) under the ground motion of 1940 El Centro earthquake (Jui-Liang et al. 2009).

responses of Case I in the x and y directions at the foundation by the choice of rigorous method are slightly different as compared with responses obtained via the approximate solutions in the current study. By comparing Fig. 6.5 and Fig. 6.6, it is noted that the translation and twist motion of the foundation of Case IV are mostly unaffected due to the small SSI effects. Moreover, the response of the roof twist of Case I obtained by both Jui-Liang et al. (2009) and current study in Fig. 6.5(c) is notably reduced as compared to that of Case IV in Fig. 6.6(c).

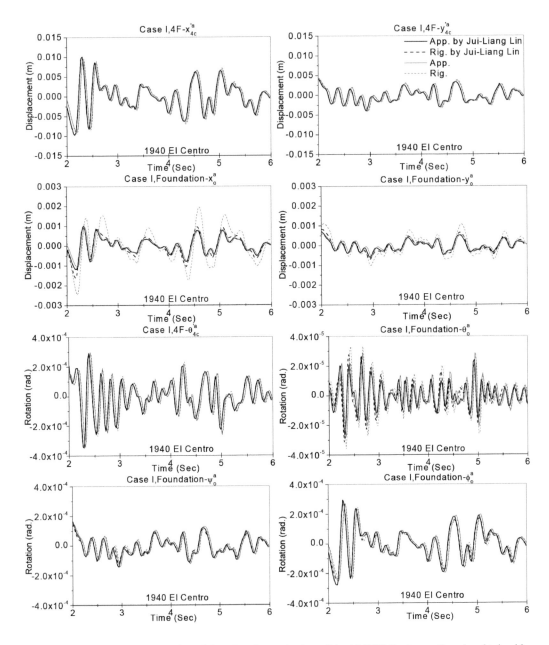

Fig. 6.5 Total response time histories of Case I under excitation of the 1940 El Centro earthquake obtained by Jui-Liang et al. (2009) and current study for both approximate and rigorous solutions.

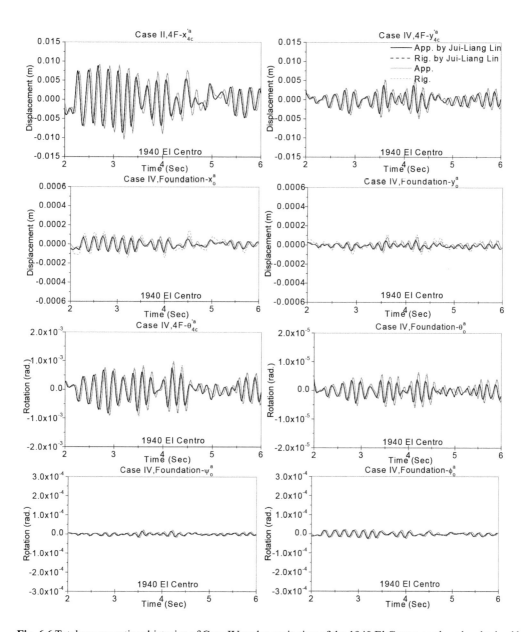

Fig. 6.6 Total response time histories of Case IV under excitation of the 1940 El Centro earthquake obtained by Jui-Liang et al. (2009) and current study for both the approximate and rigorous solutions.

6.6 Summary

As a summary of the validation of the current study, the findings of the single building using two methods are mostly consistent with the results obtained by Thambirajah et al. (1983) and Jui-Liang et al. (2009). The equation of motion of coupled buildings for the whole SSI systems using the direct integration method is solved considering the pounding effects. Although, the approximate method in the current study is validated for the building modelled by Jui-Liang et al. (2009), the approximate method used for the MDOF modal equations of motion is shown in Chapter 7 for the modal responses history analysis of the coupled buildings.

7

Case Studies

7.1 Introduction

This study examines the displacement of adjacent building structures from an analytical perspective in consideration of the effect of fluid viscous dampers. A linear model of two adjacent buildings is improved, incorporating the effects of geometric and material linearity. A three-dimensional (3D) finite element model has been defined and the linear time-history analyses have been performed to examine the seismic behaviour of the following examples. While non-linear direct integration of time history analysis can be considered in SAP 2000n computer program, linear analysis of time history is preferred to understand clearly the effect of fluid viscous damper in this application due to use of the linear parametric values of fluid viscous damper. As a result, the governing equations of motion are solved in an incremental form using Newmark's step-by-step method assuming linear variation of acceleration over a small time interval, Δt. The main aim of this study is to permit either two dynamically different buildings or the same buildings to use control forces upon one another to reduce the overall response of the system. Thereby, control systems for adjacent buildings represent a relatively new area of research that is growing rapidly (Taylor and Constantinou 1998).

The objectives of this study are:

- designing the equation of motion for adjacent buildings connected by fluid viscous dampers
- designing three-dimensional structures utilising passive control devices and designing the parameters of the devices
- developing the effectiveness of dampers when earthquake is considered in two directions
- designing the optimum placement of dampers in order to minimise the cost of dampers

7.2 Ground Motion Frequencies

All earthquake records with same time intervals were selected in order to examine the behaviour of fluid viscous damper. The earthquake time histories selected to investigate the dynamic analysis of two buildings in four example applications are: North-south (N-S) and West-east (W-E) components of Imperial Valley Irrigation District substation in El Centro, California, during the Imperial Valley, California earthquake of May, 18, 1940, N-S and W-E components of Sylmar County Hospital parking lot in Sylmar, California, during the Northridge, California earthquake of Jan. 17, 1994, N-S and W-E components of Kobe Japanese Meteorological Agency (JMA) station during the Hyogo-ken Nanbu (Kobe) earthquake of Jan. 17, 1995, N-S and W-E components of Capitola Fire Station during the Loma Prieta earthquake of Nov. 17, 1989. The peak ground acceleration values of Imperial Valley (El Centro), Kobe, Northridge and Loma Prieta earthquake motions are 0.3495, 0.8337, 0.8428 and 0.47 g, respectively (g is the acceleration due to gravity). These earthquakes have magnitudes of 7.1, 7.2, 6.8 and 6.9 respectively on the Richter scale.

All the aforementioned earthquakes have their original duration of 60s taken at a total of 3,000 time records at an interval of $\Delta t = 0.02$s. Without varying the total time number, the time interval Δt of the earthquake can be varied to alter the predominant frequency of the input motion. For example, soft soil conditions are represented by increasing the time interval while stiff or rock soil conditions occur by decreasing Δt. However, in this study, the time interval Δt is selected as 0.02s. The time history responses, including horizontal displacements, velocities, accelerations and internal forces at all joints and members in all degrees of freedom, have been computed. All the aforementioned earthquakes are shown in Fig. 7.1.

7.3 Example Buildings

For improving the dynamic behaviour of different adjacent buildings connected by dampers, four main models are presented in this application. All examples have some different characteristics. For instance, Example 1 consists of two 5-storey buildings connected by dampers as shown in Fig. 7.2(a). Example 2 has one 10-storey building and one 5-storey building where the fluid viscous dampers are placed in the floors throughout the shortest building as shown in Fig. 7.2(b). Example 3 has one 20-storey building and another 10-storey building, while Example 4 consists of two 20-storey buildings. The views of both Example 3 and Example 4 are shown in Fig. 7.2(c) and Fig. 7.2(d), respectively.

In order to investigate the effectiveness of fluid viscous, the above-mentioned examples for adjacent buildings are considered as having similar stiffness and varied stiffness. Two different cases are derived from the above examples. Case 1 indicates that two adjacent buildings are possessed of various stiffnesses, while Case 2 shows coupled buildings have similar stiffnesses. Case 1 is denominated as (a) in the examples mentioned above. For Case 2, the index of (b) is used in all the examples. For instance, Example 1(a) consists

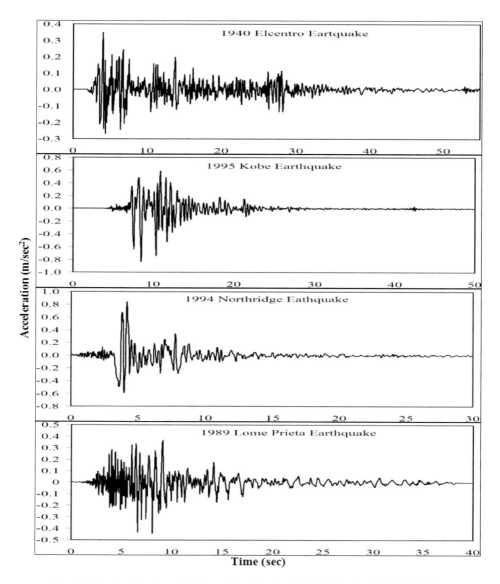

Fig. 7.1 Acceleration time histories of earthquakes in N-S direction (NOAA 2008).

of two 5-storey buildings having different shear stiffness, while Example 1(b) consists of two 5-storey buildings having similar shear stiffness. Example 2(a) has one 10-storey building and one 5-storey building which have different stiffness because of different size of columns and beams, while Example 2(b) consists of one 10-storey building and one 5-storey building having the same floor elevations with dampers linking two adjacent floors, which have different shear stiffness.

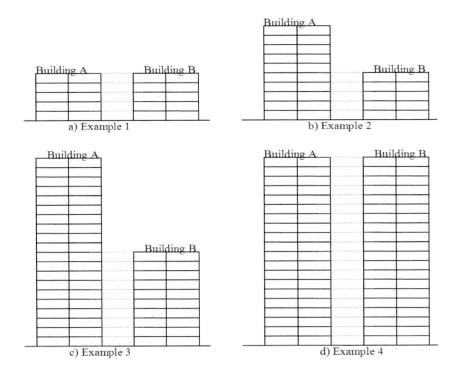

Fig. 7.2 Views of the adjacent building in four main examples.

Example 3(a) has one 20-storey building and one 10-storey building which have the dissimilar stiffness, while Example 3(b) consists of one 20-storey building and one 10-storey building having different shear stiffness. Example 4(a) consists of two 20-storey buildings having various shear stiffness. Finally, Example 4(b) consists of two 20-storey buildings having the same elevations with dampers connecting two neighbouring floors, which have the same stiffness and same structural damping ratio.

Two fluid viscous dampers are designated as Damper 1 (D1) and Damper 2 (D2) in each example. According to Xu et al. (1999), the damping coefficient was determined as 1.0×10^6 *Nsec/m* with a small variation for adjacent buildings in their studies. Therefore, for both dampers, the damping coefficients in the four main examples are determined as cd = 0.25×10^6 *Nsec/m* and cd = 0.85×10^6 *Nsec/m* respectively. The restoring force FE mentioned in Eq. 3.22 is not considered in this application in order to avoid impact load on columns and beams and to investigate the effect of damping coefficient. Hence, the damping stiffness is set at zero for joint dampers.

Details of Example 1 are explained in the following section. In order to investigate the effects of two different fluid viscous dampers on existing adjacent buildings having either different heights or same heights, four main models are presented for adjacent buildings in this application. The studies then go to the adjacent buildings consisting of one 10-storey building and one 5-storey building. One 20-storey building and one 10-storey building are examined in Example 3. Finally, analytical studies are conducted for the two 20-storey buildings in Example 4. Table 7.1 shows the sizes of columns and beams in buildings for all the examples mentioned above.

Figure 7.3 indicates the plan view of columns and beams in adjacent building for all examples, including the locations of fluid viscous dampers.

Table 7.1 The Sizes of Columns and Beams in the Buildings for Each Example.

Example	Building A			Building B		
No.	Beam Height (mm)	Beam Width (mm)	Column* Dimension (mm)	Beam Height (mm)	Beam Width (mm)	Column* Dimension (mm)
1(a)	600	250	600 × 300	500	250	500 × 300
1(b)	500	250	500 × 300	500	250	500 × 300
2(a)	600	250	600 × 300	500	250	500 × 300
2(b)	500	250	500 × 300	500	250	500 × 300
3(a)	600	300	700 × 400	500	300	600 × 300
3(b)	500	300	600 × 300	500	300	600 × 300
4(a)	600	300	700 × 400	500	300	600 × 300
4(b)	500	300	600 × 300	500	300	600 × 300

* Column dimensions are shown in Fig. 7.3

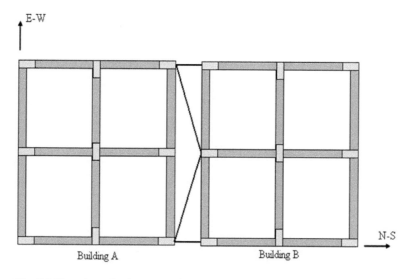

Fig. 7.3 Plan view of columns and beams in adjacent buildings for all examples.

7.3.1 *Application to Example 1*

Example 1 is composed into two different models. The primary focus of this application is shown as Example 1(a) having different shear stiffnesses. Example 1(a) consisting of two 5-storey buildings is analysed using SAP 2000n package program. Building *A* and Building *B*, which have 2-bay reinforced concrete frame, are shown in Fig. 7.4.

Adjacent buildings are connected with viscous damper devices at each storey level as shown in Fig. 7.5(a) and Fig. 7.5(b). For all modes, both buildings have damping ratios of 5 per cent of the critical structural damping ($\zeta = 0.05$). In this way, the structural damping coefficient in SAP 2000n is automatically calculated from the expression below:

$$[C] = diag\left(2M\xi\omega\right) \tag{7.1}$$

where [C] is the modal damping matrix, M, ξ and ω are the modal mass, the damping ratio and natural frequency, respectively. The mass and shear stiffness of each building are calculated. Different size of columns and beams has been used for the frames in order to investigate the sole control of fluid viscous dampers. Example 1(b) shows adjacent buildings, both having same height and same shear stiffness for matching Example 1(a). Hence, in the adjacent buildings having same heights, the importance of joint dampers can be seen on to couple buildings either having different stiffness or having same shear stiffness.

The control performances of the fluid viscous dampers are compared with both the uncontrolled adjacent structures case and the rigidly connected structures case. A thorough study is undertaken to observe the effectiveness of fluid viscous damper for multi degree of freedom adjacent buildings under various earthquake excitations. The floors apply

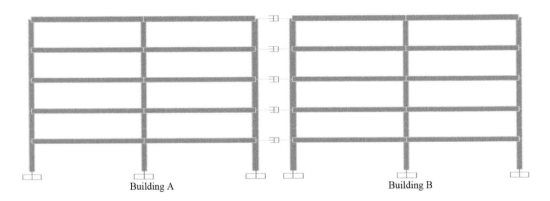

Fig. 7.4 Elevation view of the two reinforced concrete buildings for Example 1(a) and Example 1(b) in SAP 2000n computer program.

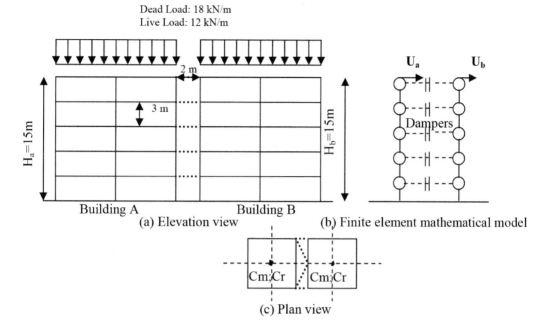

Fig. 7.5 Views of two adjacent buildings for Example 1(a) and Example 1(b).

uniformly distributed loads along the beams throughout. For Example 1(a), although the floor loads in Building *A* are same as Building *B*, the mass of Building *A* is less different than Building *B* because of the size of columns and beams. For example in 1(b), both the floor loads and mass of Building *A* are same as Building *B*. Appendices A1 and A2 show the design data of buildings and material properties. The typical slab loads at floor level of Example 1(a) and (b) are also shown in Fig. 7.5. The typical slab loads at roof level of all examples have a uniformly distributed load of 18 kN/m as dead load and a uniformly distributed load of 4.5 kN/m as live load along the beams. The typical slab loads at floor level of all examples have a uniformly distributed load of 18 kN/m as dead load and a uniformly distributed load of 12 kN/m as live load along the beams throughout. Table 7.2 shows the parameters of the structural system in buildings for all the examples mentioned above. In appendices A3, A4 and A5, slab loads are shown for each example. The distance between two adjacent buildings and the height of floors are presented as 2 m and 3 m, respectively, as shown in Fig. 7.5(a). The results in this study are demonstrated graphically in the following chapter.

7.3.2 Applications to Other Example Buildings

Studies of other buildings are conducted to find beneficial effect of fluid viscous damper for different types of adjacent buildings in order to achieve the maximum response reduction of coupled buildings under various earthquake excitations. All results of these buildings are shown separately in the following chapter. For Example 2, two different adjacent buildings are modelled as Example 2(a) and 2(b). Two adjacent buildings consisting of one 10-storey building and one 5-storey building are analysed, using SAP 2000n computer program. Building *A* is a 10-storey, 2-bay reinforced concrete frame adopted from the verification manual of the SAP 2000n package program. Building *B* is a 5-storey, 2-bay reinforced concrete frame as shown in Fig. 7.6. In Example 2(a), the buildings have different shear stiffness, although the buildings have the same shear stiffness in Example 2(b). The natural frequencies are smaller in Building *A* than Building *B* due to the two different heights of the buildings.

Figure 7.7 shows the distribution of typical slab in each storey. As shown in Fig. 7.7(b), the adjacent buildings are connected with dampers in alignment. The linked dampers at each floor have the same damping coefficient as Example 1, discussed in Section 7.3.1.

The third example, which includes one 20-storey building and one 10-storey building is analysed by using SAP 2000n package program. In Example 3, the floor load and structural damping coefficients of Building A for each storey are the same as in Building *B*. But, in Example 3(a), the shear stiffness is smaller in Building *B* than Building *A*. Example 3(b) consists of adjacent buildings having the same shear stiffness but with different heights. Hence, the structural heights of Example 2 discussed in the previous section, which include the mass and shear stiffness, are changed in Example 3 to check the effectiveness of joint

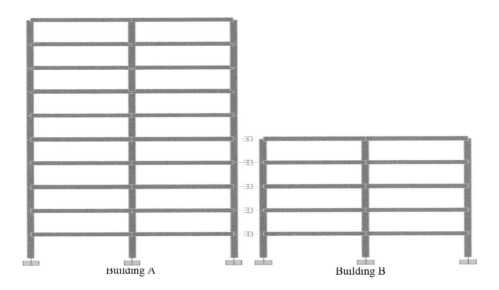

Building A   Building B

Fig. 7.6 Elevation view of the two reinforced concrete buildings for Example 2(a) and Example 2(b) in SAP 2000n computer program.

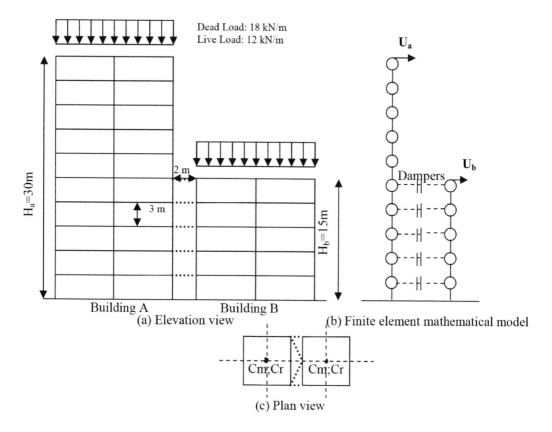

Dead Load: 18 kN/m
Live Load: 12 kN/m

(a) Elevation view

(b) Finite element mathematical model

(c) Plan view

Fig. 7.7 Views of two adjacent buildings for Example 2(a) and Example 2(b).

dampers. Figure 7.8 shows the model view of the reinforced concrete buildings with two-dimension views of the sizes of beams and columns.

Figure 7.9 shows the distribution of typical slab at each storey. As shown in Fig. 7.9(b), the adjacent buildings are connected with dampers in alignment. The linked dampers at each floor have the same stiffness and damping coefficient as discussed in Section 7.3. The centre of rigidity in the buildings (Cm) overlaps the geometric centre of gravity of the buildings (Cr) as shown in Fig. 7.9(c). Hence, the torsion effects can remain at minimum level for both the buildings. For all modes, both the buildings have damping ratios of 5 per cent of the critical structural damping ($\zeta = 0.05$) as the previous examples.

The last example consists of two parts as the previous examples. Example 4 has two 20-storey adjacent buildings with same floor elevations and dampers connecting two neighbouring floors. In this application for Example 4(a), the mass and shear stiffness

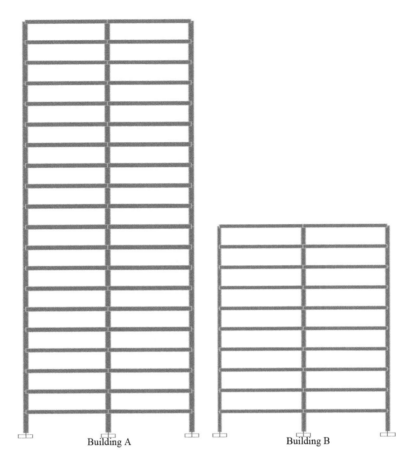

Building A   Building B

Fig. 7.8 Model view of two reinforced concrete buildings for Example 3(a) and Example 3(b) in SAP 2000n computer program.

of Building *A* are selected with different characteristics as Building *B*. Building *A* and Building *B* are two 20-storey buildings having the same floor elevations, 2-bay reinforced concrete frame as shown in Fig. 7.10 with two dimensional (2D) views of the sizes of columns and beams in each building. Example 4(a) consists of two 20-storey adjacent buildings having the same floor elevations but with different shear stiffness. The shear stiffness is smaller in Building *B* than Building *A*. Hence, the natural frequencies of the two buildings are smaller in Building *B* than Building *A*. The adjacent buildings are connected with dampers in alignment. The linked dampers at each floor have the same damping coefficient as discussed in Sections 7.3.1 and 7.3.2.

The floor mass and storey stiffness are considered to be uniform in both the buildings. Figure 7.11(a) indicates the allocation of typical slab at each storey.

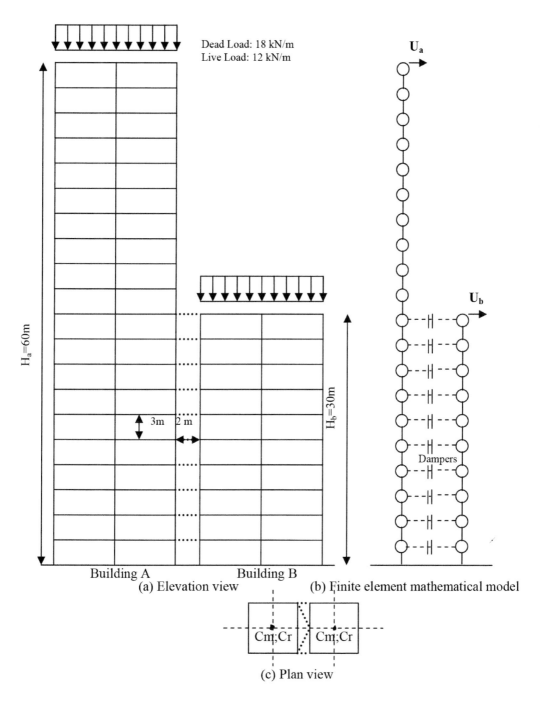

Dead Load: 18 kN/m
Live Load: 12 kN/m

Building A
Building B
(a) Elevation view
(b) Finite element mathematical model

(c) Plan view

Fig. 7.9 Views of two adjacent buildings for Example 3(a) and Example 3(b).

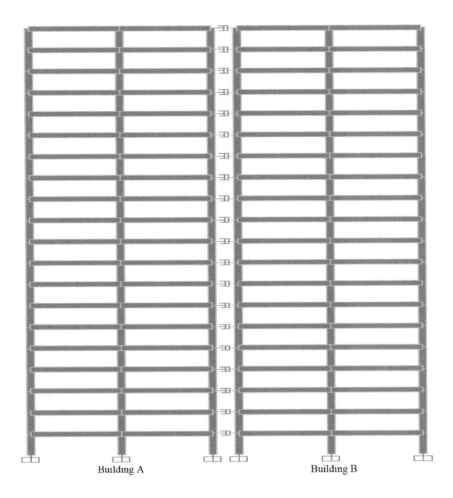

Building A   Building B

Fig. 7.10 Model view of two reinforced concrete buildings for Example 4 in SAP 2000n computer program.

The centre of rigidity of the buildings (Cm) overlaps the geometric centre of gravity of the buildings (Cr) as shown in Fig. 7.11(c). Hence, the torsion effects can remain at minimum levels for both the buildings. The masses of the two buildings are assumed to be same and the damping ratio in each building is taken as 5 per cent. In this application, the damping coefficients of fluid viscous dampers are selected with the same characteristics as the previous example. As shown in Fig. 7.11, the adjacent buildings are connected with dampers in alignment. Example 4(b) consists of two 20-storey adjacent buildings having the same floor elevations with the same shear stiffness. The typical slab loads in each storey for Example 4(b) are the same as for Example 4(a) in Fig. 7.11.

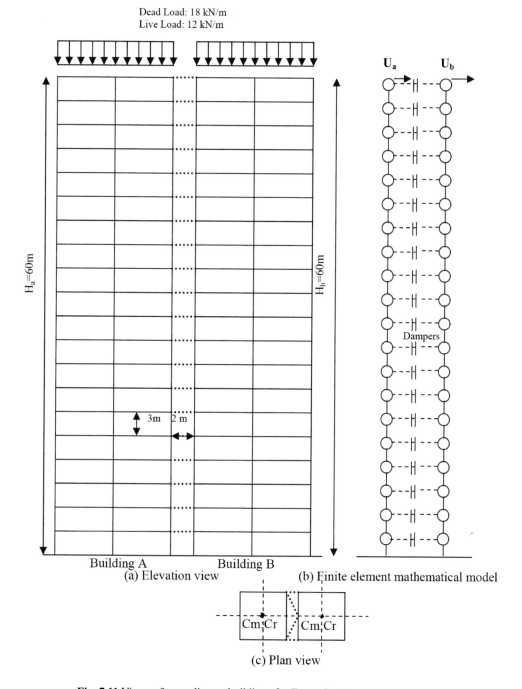

Fig. 7.11 Views of two adjacent buildings for Example 4(a) and Example 4(b).

7.4 Summary

In this chapter, four different models for adjacent buildings with either the same stiffness or different stiffness are designed by using SAP 2000n package program. The aim of this chapter is to create different types of coupled buildings in order to investigate the benefits of fluid viscous dampers. After a brief overview of earthquake time histories is presented in this chapter to examine the seismic behaviour of the two buildings in all the examples, the building models are described as frame buildings and do not include shear buildings. For all models, the damper damping coefficient remains unchanged.

The first example is two 5-storey buildings having 2-bay reinforced concrete frames. The adjacent buildings consisting of the same floor elevations are shown in this chapter. Moreover, Example 1 has two parts. In Building *A* in Example 1(a), the stiffness of the columns is bigger than Building *B*, although Building *A* in Example 1(b) is completely the same as Building *B* in terms of the dynamic characteristics. In this example, the aim is to show the overall effectiveness of the dampers in the same adjacent buildings in terms of the dynamic characteristics, but with the same heights.

The second example is one 10-storey Building *A* and one 5-storey Building *B* where each building consists of 2-bay reinforced concrete frames. Although these buildings have the same mass and structural damping coefficient, the heights of the adjacent buildings are different. In this example, the aim is to demonstrate the overall effectiveness of dampers in the same coupled buildings in terms of the dynamic characteristics, but with different heights. As Example 1, Example 2 consists of two parts in conjunction with either the same stiffness or different stiffness.

The third example is one 20-storey Building *A* and one 10-storey Building *B* that include 2-bay reinforced concrete frames. For Example 3(a), the shear stiffness of Building *A* is more than the shear stiffness of Building *B* because the widths of columns and beams in Building *A* are wider as an alteration from Example 2. Hence, the mass and shear stiffness of the buildings are different. In this example, the aim is to demonstrate the overall effectiveness of dampers in different coupled buildings in terms of dynamic characteristics, but with different heights.

Finally, Example 4 is two 20-storey reinforced concrete buildings having different stiffness in each building. The aim of Example 4(a) is to investigate the benefits of dampers in conjunction with different shear stiffness but with the same heights. Example 4(b) has two 20-storey buildings having 2-bay reinforced concrete frames. The adjacent buildings consisting of the same floor elevations are shown in this chapter. Moreover, Building *A* is completely same as Building *B* in terms of the dynamic characteristics. In this example, the aim is to show the overall effectiveness of dampers in the same adjacent buildings in terms

Table 7.2 Parameters of Structure System in the Buildings for Examples (a).

Floor No.	Example 1(a) A		Example 1(a) B		Example 2(a) A		Example 2(a) B		Example 3(a) A		Example 3(a) B		Example 4(a) A		Example 4(a) B	
	K	M	K	M	K	M	K	M	K	M	K	M	K	M	K	M
1	324	183	215	177	324	183	215	177	756	196	324	183	756	196	324	183
2	324	183	215	177	324	183	215	177	756	196	324	183	756	196	324	183
3	324	183	215	177	324	183	215	177	756	196	324	183	756	196	324	183
4	324	183	215	177	324	183	215	177	756	196	324	183	756	196	324	183
5	324	184	215	179	324	183	215	179	756	196	324	183	756	196	324	183
6	–	–	–	–	324	183	–	–	756	196	324	183	756	196	324	183
7	–	–	–	–	324	183	–	–	756	196	324	183	756	196	324	183
8	–	–	–	–	324	183	–	–	756	196	324	183	756	196	324	183
9	–	–	–	–	324	183	–	–	756	196	324	183	756	196	324	183
10	–	–	–	–	324	184	–	–	756	196	324	184	756	196	324	183
11	–	–	–	–	–	–	–	–	756	196	–	–	756	196	324	183
12	–	–	–	–	–	–	–	–	756	196	–	–	756	196	324	183
13	–	–	–	–	–	–	–	–	756	196	–	–	756	196	324	183
14	–	–	–	–	–	–	–	–	756	196	–	–	756	196	324	183
15	–	–	–	–	–	–	–	–	756	196	–	–	756	196	324	183
16	–	–	–	–	–	–	–	–	756	196	–	–	756	196	324	183
17	–	–	–	–	–	–	–	–	756	196	–	–	756	196	324	183
18	–	–	–	–	–	–	–	–	756	196	–	–	756	196	324	183
19	–	–	–	–	–	–	–	–	756	196	–	–	756	196	324	183
20	–	–	–	–	–	–	–	–	756	193	–	–	756	193	324	184
Period	$T_1 = 0.43$		$T_2 = 0.41$		$T_1 = 0.79$		$T_2 = 0.41$		$T_1 = 1.87$		$T_2 = 0.79$		$T_1 = 1.87$		$T_2 = 1.56$	

K: Storey Stiffness (10^3 kN/m)   A: Building A

M: Floor Mass (Tonne)   B: Building B

of the dynamic characteristics, but with the same heights. Table 7.2 and Table 7.3 show the parameters of the structural system in buildings for all examples mentioned above.

In all the examples, the optimum parameters of fluid viscous dampers in previous studies are used for characteristics of the dampers. For both dampers, the damping coefficients in all the four cases are determined as cd = 0.25 × 10^6 *Nsec/m* and cd = 0.85×10^6 *Nsec/m* respectively. The following chapter shows the results of these four examples to explain the effectiveness of fluid viscous damper for different types of adjacent buildings.

Table 7.3 Parameters of Structure System in the Buildings for Examples (b).

Floor No.	Example 1(b) A		B		Example 2(b) A		B		Example 3(b) A		B		Example 4(b) A		B	
	K	M	K	M	K	M	K	M	K	M	K	M	K	M	K	M
1	215	177	215	177	215	177	215	177	324	183	324	183	324	183	324	183
2	215	177	215	177	215	177	215	177	324	183	324	183	324	183	324	183
3	215	177	215	177	215	177	215	177	324	183	324	183	324	183	324	183
4	215	177	215	177	215	177	215	177	324	183	324	183	324	183	324	183
5	215	179	215	179	215	177	215	179	324	183	324	183	324	183	324	183
6	–	–	–	–	215	177	–	–	324	183	324	183	324	183	324	183
7	–	–	–	–	215	177	–	–	324	183	324	183	324	183	324	183
8	–	–	–	–	215	177	–	–	324	183	324	183	324	183	324	183
9	–	–	–	–	215	177	–	–	324	183	324	183	324	183	324	183
10	–	–	–	–	215	179	–	–	324	183	324	184	324	183	324	183
11	–	–	–	–	–	–	–	–	324	183	–	–	324	183	324	183
12	–	–	–	–	–	–	–	–	324	183	–	–	324	183	324	183
13	–	–	–	–	–	–	–	–	324	183	–	–	324	183	324	183
14	–	–	–	–	–	–	–	–	324	183	–	–	324	183	324	183
15	–	–	–	–	–	–	–	–	324	183	–	–	324	183	324	183
16	–	–	–	–	–	–	–	–	324	183	–	–	324	183	324	183
17	–	–	–	–	–	–	–	–	324	183	–	–	324	183	324	183
18	–	–	–	–	–	–	–	–	324	183	–	–	324	183	324	183
19	–	–	–	–	–	–	–	–	324	183	–	–	324	183	324	183
20	–	–	–	–	–	–	–	–	324	184	–	–	324	184	324	184
Period	$T_1 = 0.41$		$T_2 = 0.41$		$T_1 = 0.89$		$T_2 = 0.41$		$T_1 = 1.97$		$T_2 = 0.79$		$T_1 = 1.56$		$T_2 = 1.56$	

K: Storey Stiffness (10^3 kN/m)   A: Building A

M: Floor Mass (Tonne)   B: Building B

8

Results in Frequency and Time Domains

8.1 Introduction

Herein a numerical study is carried out in two sections. All the obtained results are evaluated by SAP 2000n computer program, using both frequency domain and time domain. This chapter presents the effectiveness of fluid viscous dampers investigated by way of reduction in displacement, acceleration and shear force responses of the coupled buildings in four different examples. All the results are shown with graphics taken from SAP 2000n package program in the following sections. Moreover, optimum placement of dampers for all examples is determined, creating some cases on linking dampers in the following chapter.

8.2 Results in Frequency Domain

The first section of the numerical study is that the response spectrum curves are used for the response analysis of the earthquakes in Section 4.2. In frequency domain, SAP 2000n computer program gives graphic results based on displacement-frequency and acceleration-frequency. In this section, graphs for the examples which show the displacement-frequency and acceleration-frequency are presented separately for each example.

For Example 1(a) Fig. 8.1 indicates the top floor displacement spectral density functions of the two buildings relative to the ground with and without joint dampers. As mentioned above, two different damping coefficients for joint dampers are used for each example as Damper 1 (D1) and Damper 2 (D2). With the spectral density of the unlinked buildings, the first two natural frequencies of Building A can be identified in conjunction with the related earthquakes. For example, the first two natural frequencies of Building A during the 1989 Loma Pricta earthquake in N-S direction are defined to be 3.20 and 6.82 Hz. The third natural frequency is beyond 30 Hz for Building A. The first three natural frequencies of

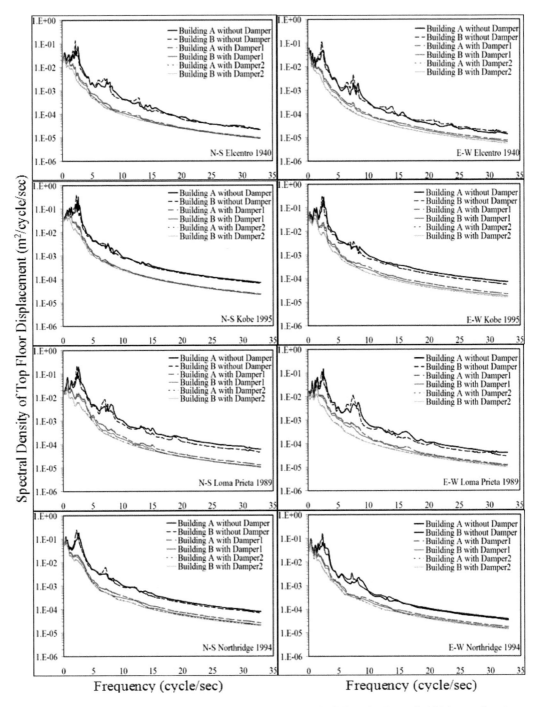

Fig. 8.1 Spectral density of top floor displacements of two adjacent buildings for Example 1(a) in two directions.

Building *B* are determined as 2.69, 6.67 and beyond 30 Hz, respectively. It is clearly seen from Fig. 8.1 that the displacement peaks of two unconnected adjacent buildings become smaller with increasing natural frequency.

In two adjacent buildings connected by jointed dampers, the first natural frequency of both Building *A* and Building *B* remain constant although the spectral density of the top floor displacement of both buildings reduces significantly in both directions. The spectra density of the top floor displacement of adjacent buildings linking Damper 2 is smaller than adjacent buildings linking Damper 1. There are no big differences in the lowest natural frequencies for both the buildings. However, the second frequency of Building *A* linking Damper 1 is slightly decreased to 6 Hz. It is seen during the 1989 Loma Prieta earthquake from Fig. 8.1 that all displacements in E-W direction are reduced significantly, installing joint dampers with optimum parameters.

Moreover, the spectral density of top floor acceleration for the buildings is shown in Fig. 8.2. The peaks in the spectral density graphs for both the buildings, which are not connected, become nearly the same with increasing natural frequency in all earthquakes, expect the 1995 Kobe earthquake. During this earthquake, the peaks became smaller with increasing natural frequency. This indicates that the contribution of higher modes of vibration to the acceleration responses can be very important for uncontrolled buildings under select earthquake movement. Newton's Second Law of Motion confirms the contribution of vibration to the acceleration responses. Additionally, the instalment of fluid viscous dampers to link two adjacent buildings indicates that peaks are significantly less, particularly at higher natural frequencies.

In Example 1(b), the top floor displacement of Building *A* without damper is the same as Building *B* without damper because both the structures are of same height and same shear stiffness as shown in Fig. 8.3.

The values of spectral density of top floor displacement for both the buildings linking dampers are changed slightly with increasing natural frequency, although both buildings have the same characteristics. However, in E-W direction, there is no big difference between adjacent buildings without damper and adjacent building with damper. Figure 8.4 indicates the spectral density of top floor acceleration of adjacent buildings for Example 1(b) in N-S and E-W directions.

It can be seen from Fig. 8.4 that the peaks of top floor accelerations for Building *B* with Damper 1 become smaller than Building *A* with Damper 1. In Example 2(a), Fig. 8.5 shows the top floor displacement spectral density functions of coupled buildings in terms of being with and without joint dampers. According to the spectral density of buildings in Fig. 8.5, the first three natural frequencies can be clearly seen in association with the related earthquake.

In Building *A* without dampers in N-S Loma Prieta 1989 earthquake, the first three natural frequencies are identified as 0.9, 2.7, 6.8 Hz respectively, while the first three natural

129

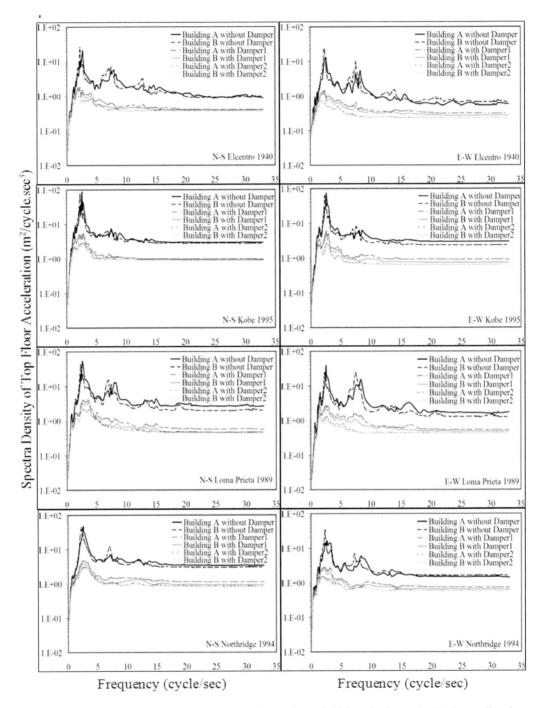

Fig. 8.2 Spectral density of top floor acceleration of two adjacent buildings for Example 1(a) in two directions.

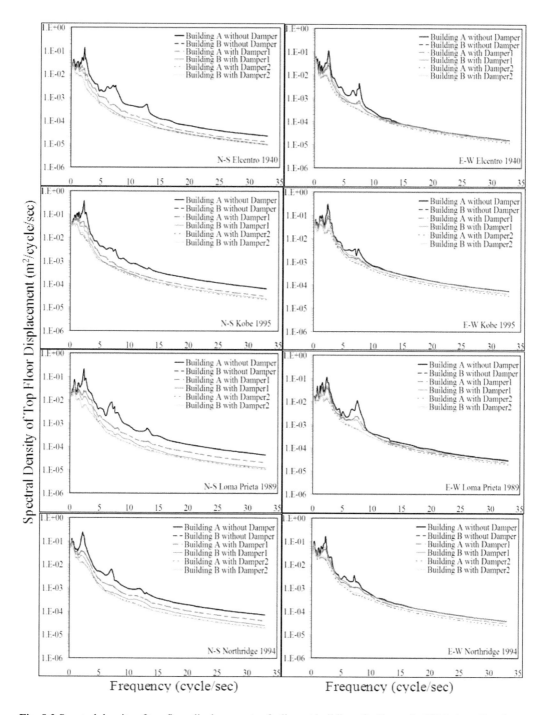

Fig. 8.3 Spectral density of top floor displacements of adjacent buildings for Example 1(b) in two directions.

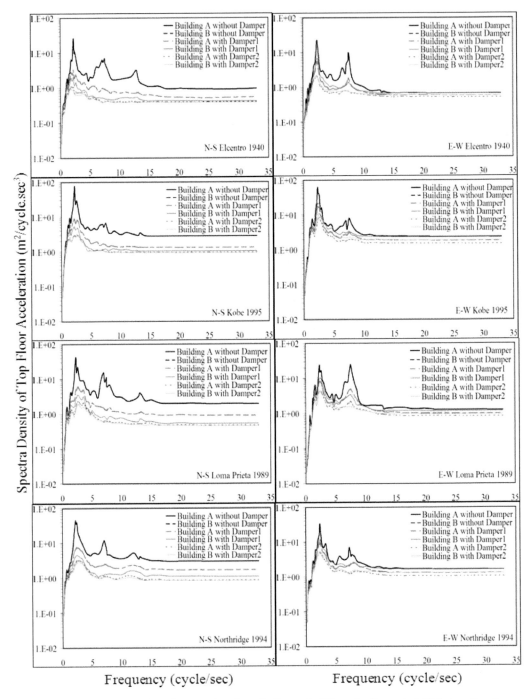

Fig. 8.4 Spectral density of top floor acceleration of adjacent buildings for Example 1(b) in two directions.

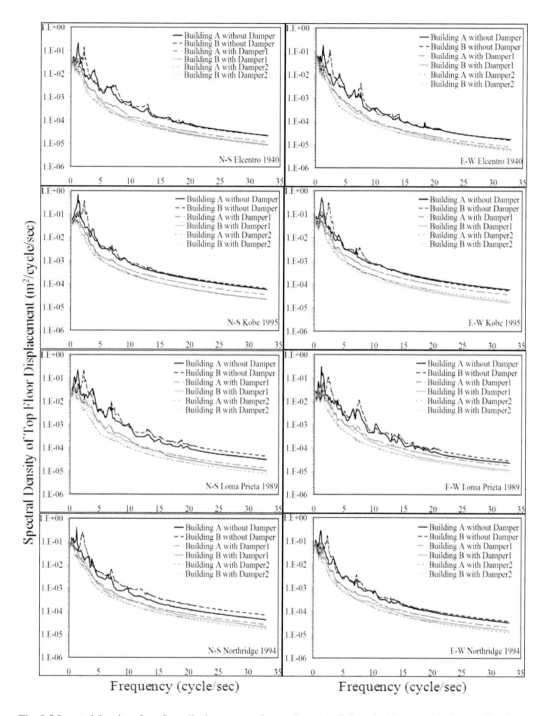

Fig. 8.5 Spectral density of top floor displacements of two adjacent buildings for Example 2(a) in two directions.

frequencies of Building *A* with Damper 1 are determined as 1.1, 2.8, 7 Hz respectively. For Building *B* without dampers in N-S Loma Prieta 1989 earthquake, the first three natural frequencies are found to be 1.1, 3.6, 6.5 Hz, while these frequencies for Building *B* connected by jointed Damper 1 are determined as 1, 2.9, 6.4 Hz respectively. These frequencies clearly show that the modes of the buildings are well separated. The displacements in the lowest natural frequencies for Building *B* connected by Damper 1 and Damper 2 become smaller with increasing frequency in all the earthquakes when it is compared with Building *B* unconnected by dampers. In addition, all displacements of Building *B* are reduced significantly, installing joint dampers with optimum parameters. As discussed in Example 1(a), it can be seen from Fig. 8.1 and Fig. 8.5 that the dampers are more effective for lower buildings than the higher ones in terms of the reduction of displacements in both the lowest and highest frequencies.

The spectral density of the top floor acceleration for buildings is indicated in Fig. 8.6. Although the peaks in these graphs for Building *A* become smaller with increasing frequencies, the peaks for Building *B* become higher significantly in the 1989 Loma Prieta earthquake and 1994 Northridge earthquake for both the directions.

However, the reduction of peaks becomes higher after using joint dampers for Building *B*. As a result, it is shown in Example 2(a) that in coupled buildings with different heights, the effectiveness of fluid viscous dampers become less important for medium-rise building than the low-rise buildings.

In Example 2(b), Fig. 8.7 shows the spectral density of top floor displacements of adjacent buildings for Example 2(b) in two directions. Generally, the linked adjacent buildings indicate that peaks during all earthquakes at higher natural frequencies are replaced slowly with increasing natural frequency.

It can be seen in Fig. 8.7 that although both buildings have the same characteristics in terms of shear stiffness and structural damping ratio, the amount of reduction in the peaks for Building *B* is more than Building *A* due to having different heights. Figure 8.8 investigates the spectral density of top floor acceleration of adjacent buildings for Example 2(b) in two directions.

It can be said in terms of acceleration that Damper 2 is more effective for the decreasing than Damper 1. Placement of dampers as diagonals becomes important to provide the reduction in the E-W direction. For Example 3(a), the results of the top floor displacement spectral density functions and the top floor acceleration functions of both the 20-storey building and the 10-storey building are as shown in Fig. 8.9 and Fig. 8.10.

The reduction of displacements of Building *B* is more obvious than Building *A*. Hence, it can be observed that the fluid joint dampers can be more effective for a low-rise Building *B* than the high-rise Building *A*.

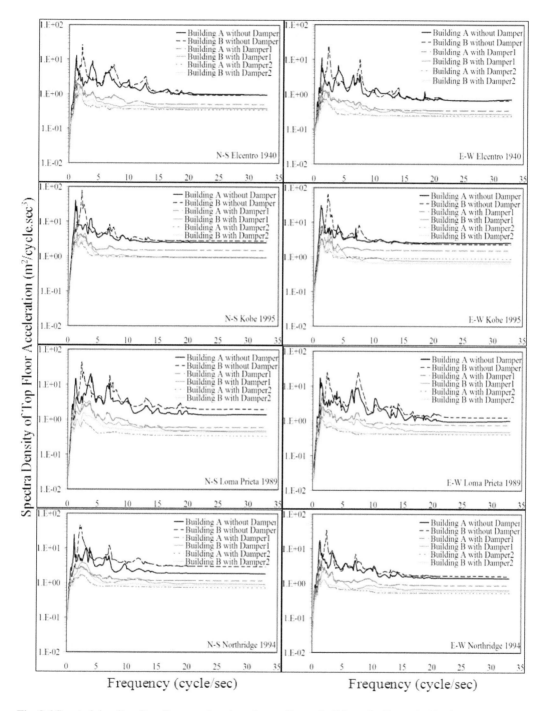

Fig. 8.6 Spectral density of top floor acceleration of two adjacent buildings for Example 2(a) in two directions.

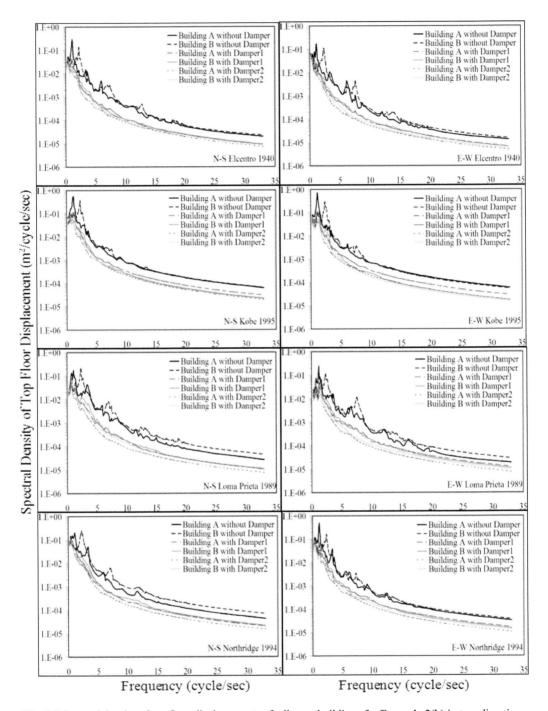

Fig. 8.7 Spectral density of top floor displacements of adjacent buildings for Example 2(b) in two directions.

136

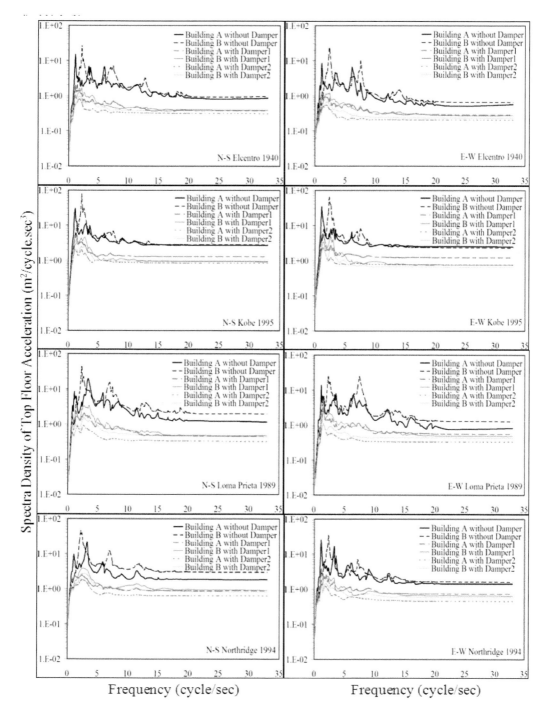

Fig. 8.8 Spectral density of top floor acceleration of adjacent buildings for Example 2(b) in two directions.

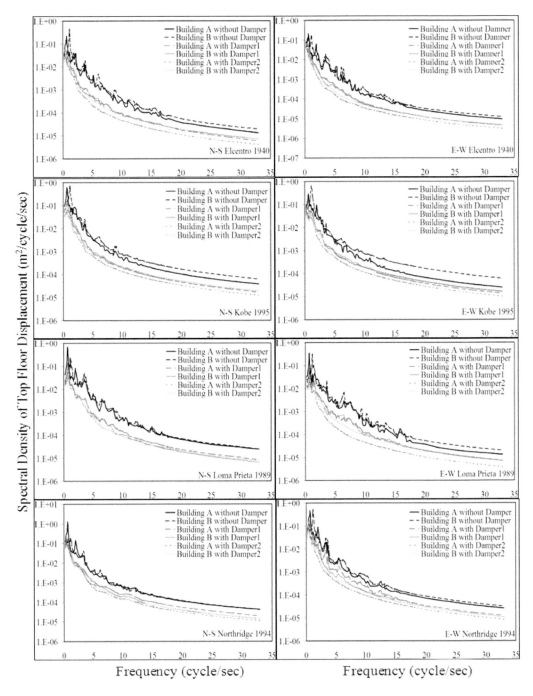

Fig. 8.9 Spectral density of top floor displacements of two adjacent buildings for Example 3(a) in two directions.

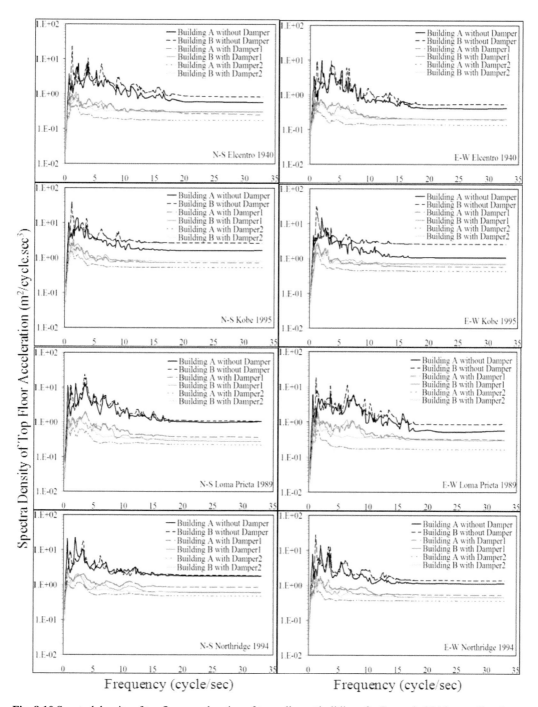

Fig. 8.10 Spectral density of top floor acceleration of two adjacent buildings for Example 3(a) in two directions.

It can be noted from Fig. 8.10 that the peaks of accelerations occur often in the lowest frequencies for either adjacent buildings linking dampers or coupled buildings without dampers. It is interesting that although Damper 2 has more damping coefficient than Damper 1 and medium-rise buildings linking Damper 2 are more effective in terms of the reduction of acceleration than high-rise buildings linking Damper 2, the amount of the reduction of acceleration for both the buildings linking Damper 2 become similar to both the buildings connecting Damper 1 in the E-W direction. For Example 3(b), Fig. 8.11 shows spectral density of top floor displacements of the adjacent buildings, for Example 3(b), in two directions.

As shown in Fig. 8.11, when adjacent buildings having the same characteristics but different heights are connected with either Damper 1 or Damper 2, the first three natural frequencies are almost similar on the adjacent buildings without related dampers even though the peaks become smaller with increasing frequencies. Figure 8.12 indicates spectral density of top floor acceleration of adjacent buildings for Example 3(b) in two directions. It can be seen from Fig. 8.12 that the peaks show more often the increasing frequencies. Therefore, increase of the number of storeys between adjacent buildings can cause more peaks of adjacent buildings.

With reference to Example 4(a), the results of the top floor displacement spectral density functions for both the buildings are shown in Fig. 8.13. It can be seen that the buildings have small differences in the spectral density functions of the top floor displacements in the highest frequencies, especially in N-S Kobe 1995 and N-S Northridge earthquakes. The peaks in Fig. 8.13 for adjacent buildings become smaller with increasing natural frequencies during all the earthquakes. In Building A without dampers in N-S Loma Prieta 1989, the first two natural frequencies are identified as 0.6, 1.8 Hz respectively, while the first two natural frequencies of Building A with Damper 1 are determined as 0.8, 2.0 Hz respectively. For Building B without dampers in N-S Loma Prieta 1989 earthquake, the first two natural frequencies are found to be 0.6, 1.5 Hz, while frequencies for Building B connected by jointed dampers are determined as 0.7, 1.6 Hz respectively.

The spectral density of the top floor acceleration for buildings is shown in Fig. 8.14. Although the peaks in these graphs for unlinked buildings become smaller with increasing frequencies, the peaks for linked buildings become smaller significantly. As a result, it is shown in Example 1 and Example 4 that the effectiveness of fluid viscous dampers becomes less important for high-rise adjacent buildings than low-rise adjacent buildings.

It can be noted from Fig. 8.15 and Fig. 8.16 that both the buildings show the same results because of having same characteristics. There are no big differences in the highest and lowest natural frequencies for both the linked and unlinked buildings. Moreover, the peaks in Fig. 8.16 for adjacent buildings become smaller in the highest natural frequencies during all earthquakes. The peaks become nearly the same as in the highest frequencies. The peaks for both buildings remain constant in the lowest frequencies expect the N-S Loma

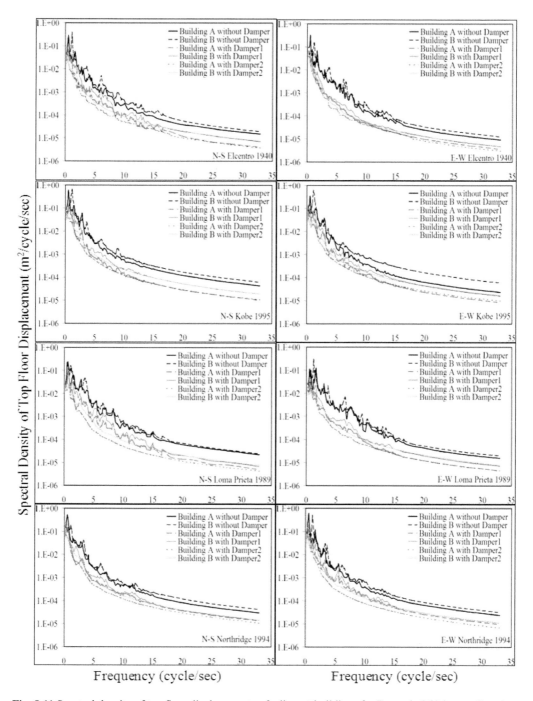

Fig. 8.11 Spectral density of top floor displacements of adjacent buildings for Example 3(b) in two directions.

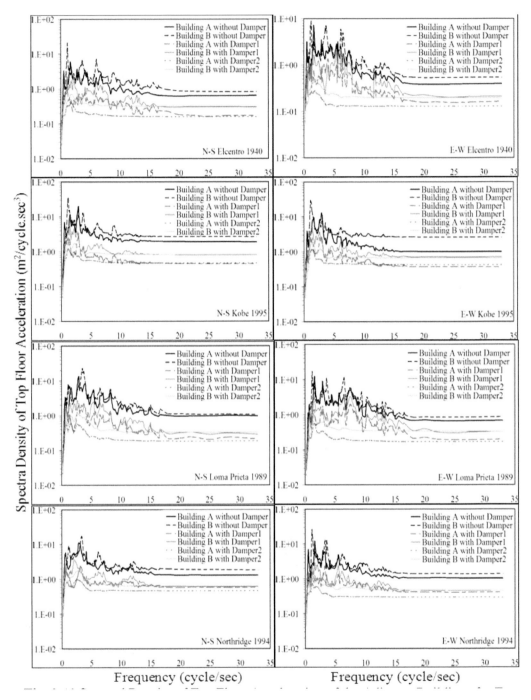

Fig. 8.12 Spectral density of top floor acceleration of adjacent buildings for Example 3(b) in two directions.

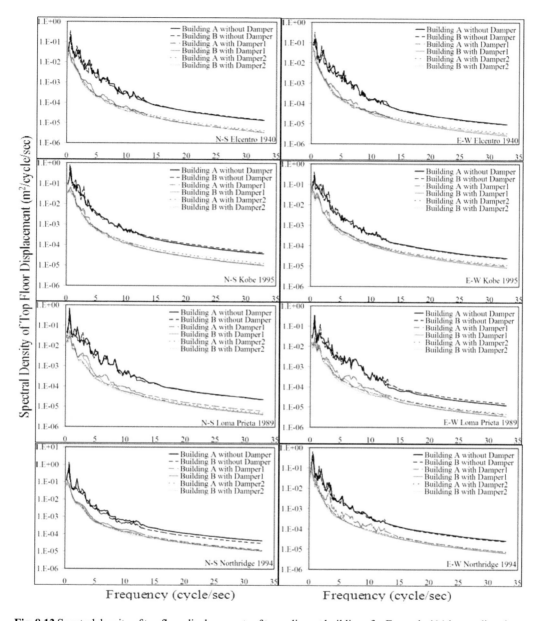

Fig. 8.13 Spectral density of top floor displacements of two adjacent buildings for Example 4(a) in two directions.

Prieta 1989 earthquake. During this earthquake, there is no effect of dampers on adjacent buildings in the lowest frequencies.

Moreover, in contrast to Example 4(b), the effectiveness of fluid viscous dampers in Example 4(a) can be clearly seen in the lowest frequencies in N-S Northridge 1994.

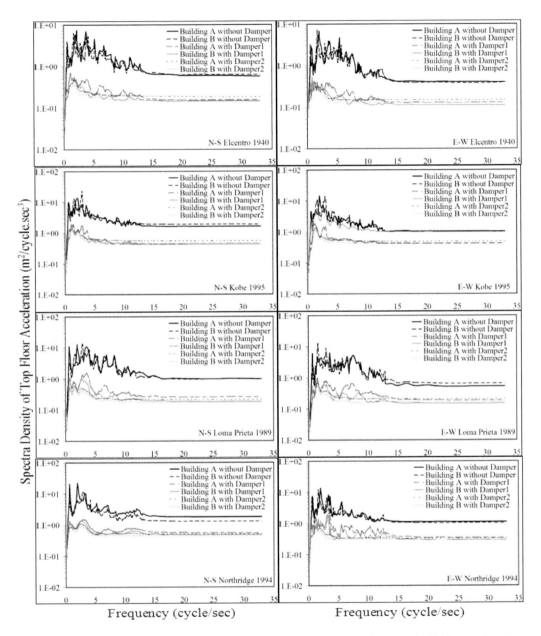

Fig. 8.14 Spectral density of top floor acceleration of two adjacent buildings for Example 4(a) in two directions.

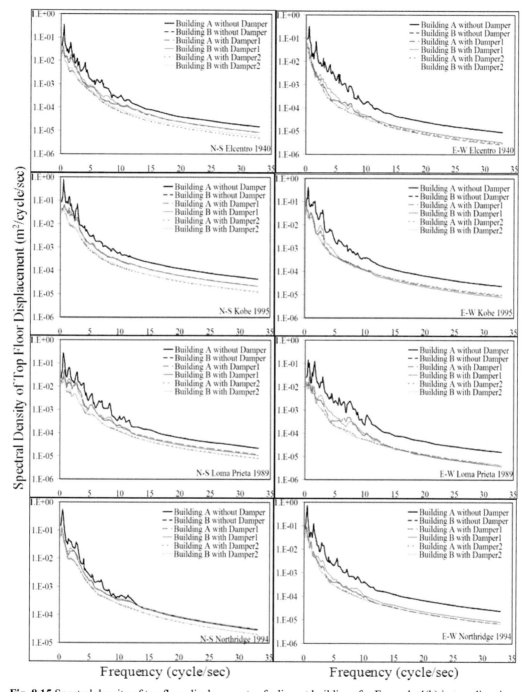

Fig. 8.15 Spectral density of top floor displacements of adjacent buildings for Example 4(b) in two directions.

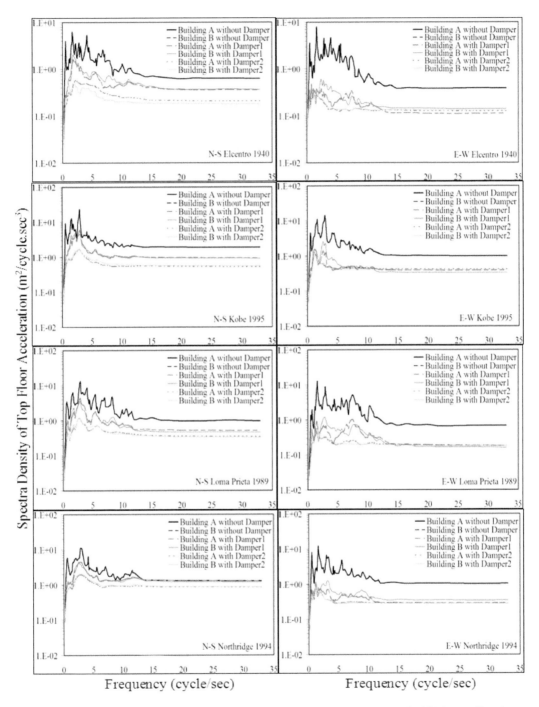

Fig. 8.16 Spectral density of top floor acceleration of adjacent buildings for Example 4(b) in two directions.

8.3 Results in Time Domain

In the second section of the numerical study, the graphs of displacement-time and shear force-time are presented with the results obtained from SAP 2000n package program. In time domain, the graphs of the examples which show the displacement-time and the shear force-time are presented separately for each example to confirm the effectiveness of joint dampers.

While examining the displacements of seismic response, the coupled building structures in all the examples are subjected to earthquake ground motion with time history of ground acceleration of four simulated earthquakes, which are derived from: (i) *El Centro*. The N-S and E-W component recorded at the Imperial Valley Irrigation District Substation in El Centro, California, during the Imperial Valley, California earthquake of May 18, 1940. (ii) *Northridge*. The N-S and E-W component recorded at Sylmar County Hospital Parking Lot in Sylmar, California, during the Northridge, California earthquake of January 17, 1994. (iii) *Kobe*. The N-S and W-E component recorded at Kobe Japanese Meteorological Agency (JMA) Station, during the Hyogo-ken Nanbu, Kobe earthquake of January 17, 1995. (iv) *Loma Prieta*. The N-S and W-E component recorded at Capitola Fire Station, during the Loma Prieta earthquake of November 17, 1989. For examining the shear force, buildings have been subjected to earthquake ground motion with time history of ground acceleration by the earthquakes mentioned above. This study is shown with graphs, including the two directions. The graphs of shear force-time for all examples are shown below. In Example 1, the time histories of the top floor displacement of adjacent buildings in N-S direction are presented in Fig. 8.17, respectively, with and without the joint dampers.

It should be noted that within the first three seconds, the amplitudes of displacement of both buildings are not reduced. However, between the first three seconds to ten seconds, the peak responses of both the buildings in N-S 1989 Loma Prieta earthquake are reduced with the peak response reduction range from 45–65 per cent for adjacent buildings linking Damper 1.

Figure 8.18 displays time histories of top floor displacement of adjacent buildings for Example 1(a) in E-W direction. It is interesting that the top floor displacements for Building *A* connecting Damper 2 are reduced during peak response reduction range at around 50 per cent. The reduction range for Building *B* linking Damper 2 is 65 per cent after first five seconds, due to the fact that the shear stiffness is smaller in Building *B* than Building *A*.

Figure 8.19 shows shear force-time graphs for Example 1(a) in two directions. The time histories of the base shear force responses in E-W direction to find the effective behaviour of damper on shear force of the floors, by decreasing the values of force. It is interesting that the amplitude of shear force for Building *B* is not reduced for both Damper 1 and Damper 2 in 1940 Elcentro and 1995 Kobe earthquakes in E-W direction, although the amplitude of shear force for Building *A* are reduced with the peak response reduction range from 10–52% within the first three seconds of these earthquakes.

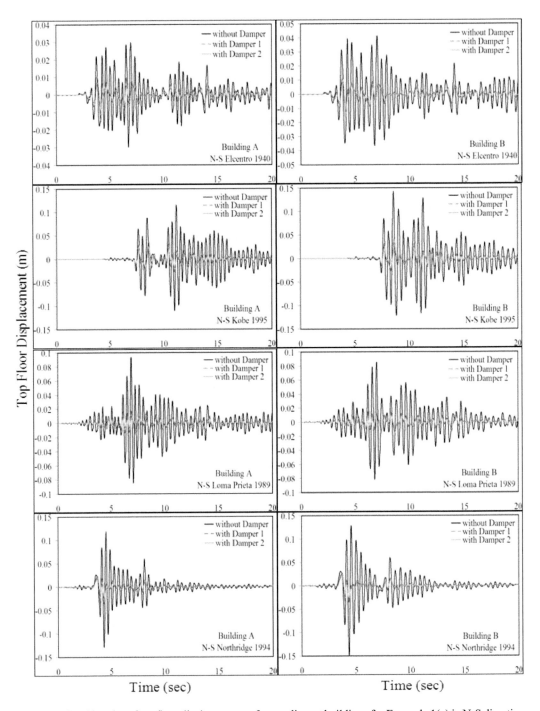

Fig. 8.17 Time histories of top floor displacements of two adjacent buildings for Example 1(a) in N-S direction.

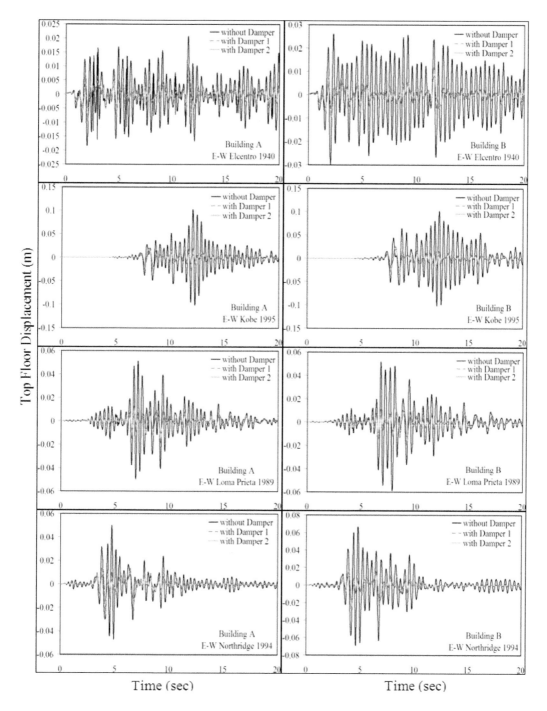

Fig. 8.18 Time histories of top floor displacement of adjacent buildings for Example 1(a) in E-W direction.

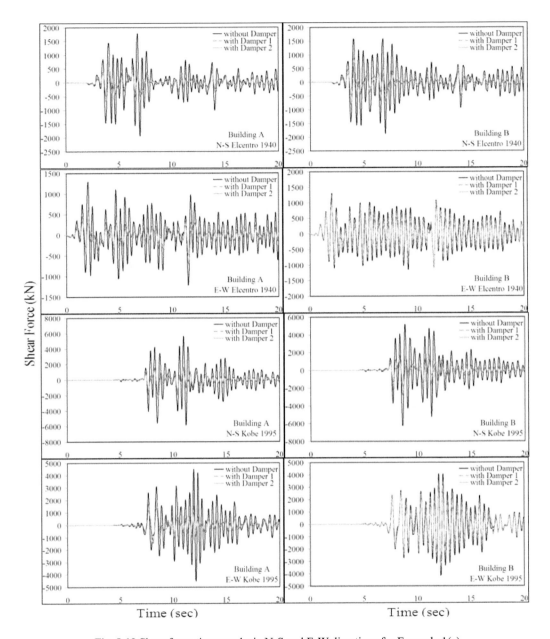

Fig. 8.19 Shear force-time graphs in N-S and E-W directions for Example 1(a).

For Example 1(b), Fig. 8.20 indicates time histories of top floor displacement of adjacent buildings in two directions. As expected, both the buildings either linking Damper 1 or connecting Damper 2 have the same reduction range in two directions owing to their same characteristics. For this reason, Fig. 8.20 investigates the displacements in two directions.

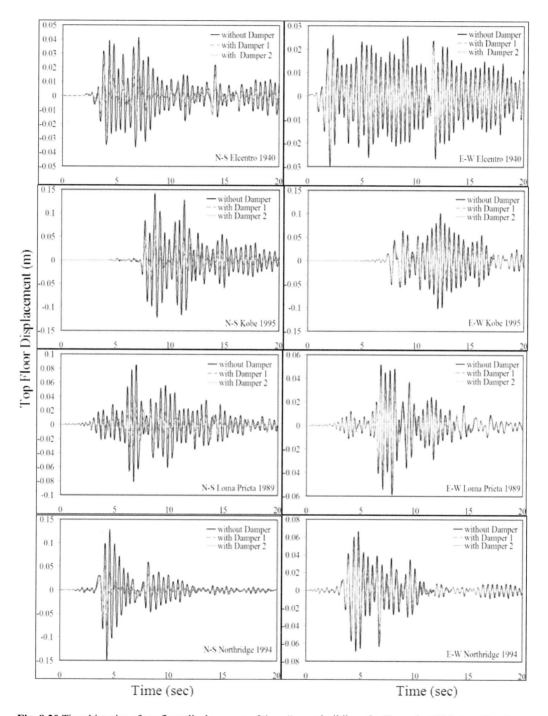

Fig. 8.20 Time histories of top floor displacement of the adjacent buildings for Example 1(b) in two directions.

It can be clearly seen that the amplitudes of top floor displacements of adjacent buildings with same shear stiffness and same height in E-W direction are not reduced with Damper 1 and Damper 2.

Figure 8.21 shows shear force-time graphs of adjacent buildings for Example 1(b) in two directions. As mentioned above, for the same adjacent buildings, the effectiveness of dampers can be seen in N-S direction in terms of the reduction of shear force, while the efficacy of dampers for both buildings in E-W direction cannot be seen in Fig. 8.21.

In Example 2(a), Fig. 8.22 indicates the time histories of the top floor displacement of adjacent buildings in N-S direction, respectively, with and without the related dampers. Within the first nine or eleven seconds, the amplitudes of displacement of Building *A* are reduced with the peak response reduction ranging between 20–45 per cent with Damper 1 while the peak response reduction range is between 25–65 per cent with Damper 2 in N-S Elcentro 1940 earthquake. However, after the first eleven seconds, the peak responses of Building *B* in N-S direction are reduced with the peak response reduction ranging between 30–70 per cent.

Time histories of top floor displacement of adjacent buildings for Example 2(a) in E-W direction are shown in Fig. 8.23. It can be said that the reductions of top floor displacement are changed slightly in Building *A*, whereas the peak response reduction range changes significantly.

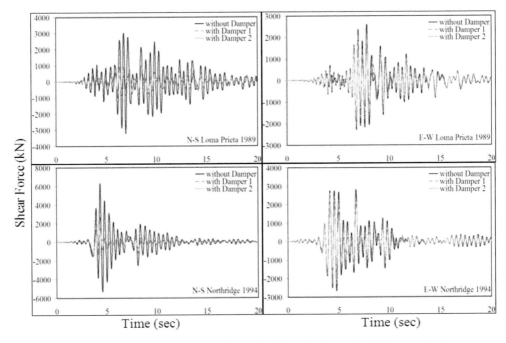

Fig. 8.21 Shear force-time graphs of adjacent buildings for Example 1(b) in two directions.

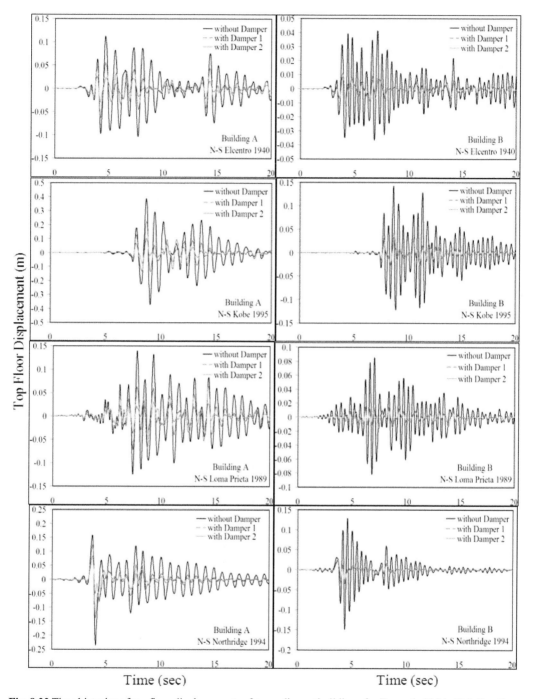

Fig. 8.22 Time histories of top floor displacements of two adjacent buildings for Example 2(a) in N-S direction.

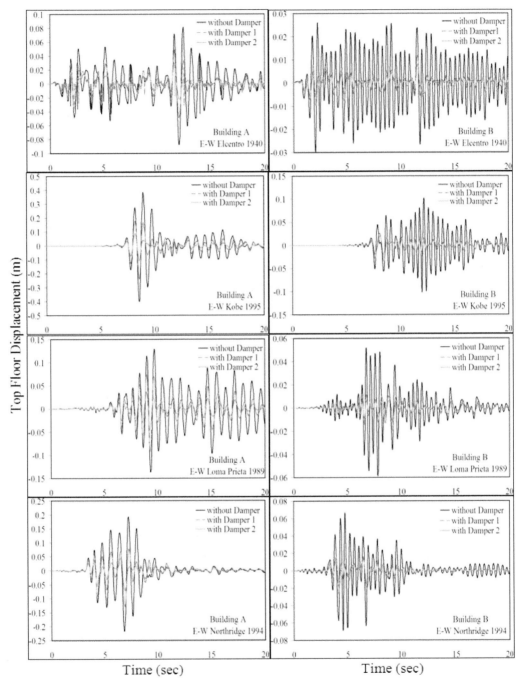

Fig. 8.23 Time histories of top floor displacement of adjacent buildings for Example 2(a) in E-W direction.

For investigating the benefits of fluid viscous dampers on shear forces of buildings, SAP 2000n computer program shows the shear force-time graphs together for adjacent buildings. Figure 8.24 shows the time histories of the base shear force responses for both Northridge 1994 and Loma Prieta 1989 in two directions, with decreasing values of force. It is seen that the force responses of Building *A* do not reduce significantly for all directions.

Time histories of top floor displacement of adjacent buildings for Example 2(b) in N-S direction are examined in Fig. 8.25. The amplitude of reduction of displacement for Building *B* with Damper 2 is higher than the peak response reduction range of Building *A* with Damper 2 in N-S Northridge 1994.

In E-W direction, when Example 2(b) compares with Example 1(b), it is seen that Building *B* which is a 5-storey building, has greater reduction ranges in terms of displacement as seen in Fig. 8.26. Therefore, using dampers for adjacent buildings with different heights is more beneficial than adjacent buildings having similar heights.

Figure 8.27 indicates the shear force-time graphs of adjacent buildings for Example 2(b) in two directions. It is seen that there are not any differences in amplitudes of displacement for E-W direction.

For Example 3(a), Fig. 8.28 demonstrates the time histories of top floor displacements of two adjacent buildings for Example 3(a) in N-S direction. The adjacent buildings having different dynamic characteristics and are connected by dampers in N-S direction. It can be observed from N-S Northridge 1994 in Fig. 8.28 that both stiffness and heights of Building *A* are higher than Building *B*. For this reason, reduction of top floor displacements of Building *A* with either Damper 1 or Damper 2 is less than that of Building *B* with either Damper 1 or Damper 2.

Another interesting observation in Fig. 8.28 is use of Damper 1 or Damper 2 in Building *B* does not change in terms of the reduction rate of displacements of Building *B* without dampers. Figure 8.29 shows that displacements of Building *A* linking Damper 1 are similar to Building *A* without dampers in E-W direction, except for 1989 Loma Prieta and 1994 Northridge earthquakes.

Figure 8.30 shows shear force-time graphs in N-S and E-W directions for Example 3(a). In N-S direction, the shear forces of Building *A* are changed significantly according to time when it is compared with E-W direction.

For Example 3(b), time histories of top floor displacement of adjacent buildings for Example 3(b) in N-S direction are shown in Fig. 8.31. In N-S Elcentro 1940 and N-S Kobe 1995, the values of displacements for Building *A* linking Damper 1 change with peak reduction ranging between 10–50 per cent within the first ten seconds, while peaks for Building *A* connecting Damper 2 reduce from 20–70 per cent within the first ten seconds.

Moreover, using dampers in terms of reduction range for Building *B* is more important than Building *A*. Figure 8.32 investigates the time histories of top floor displacement of adjacent buildings, for Example 3(b) in E-W direction.

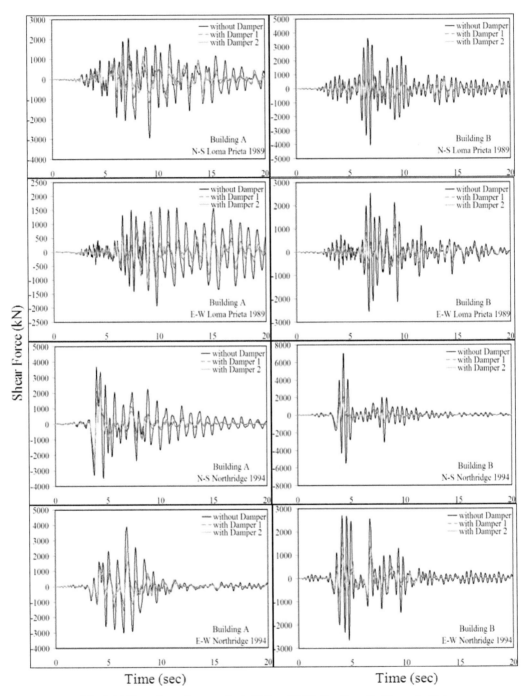

Fig. 8.24 Shear force-time graphs in N-S and E-W directions for Example 2(a).

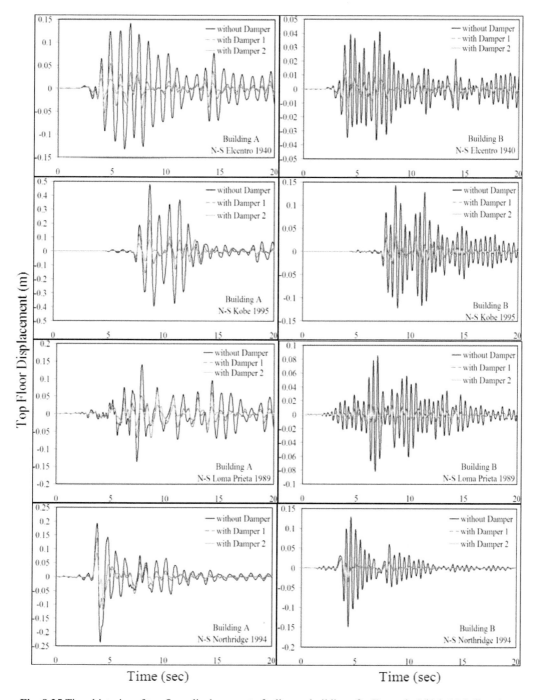

Fig. 8.25 Time histories of top floor displacement of adjacent buildings for Example 2(b) in N-S direction.

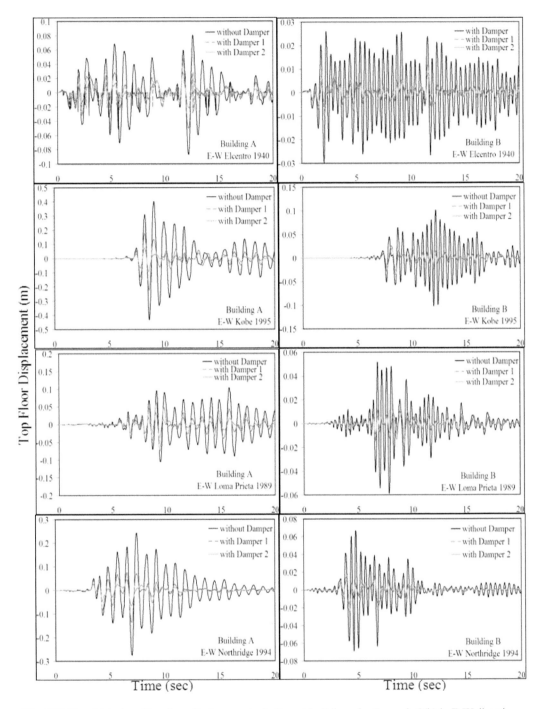

Fig. 8.26 Time histories of top floor displacement of adjacent buildings for Example 2(b) in E-W direction.

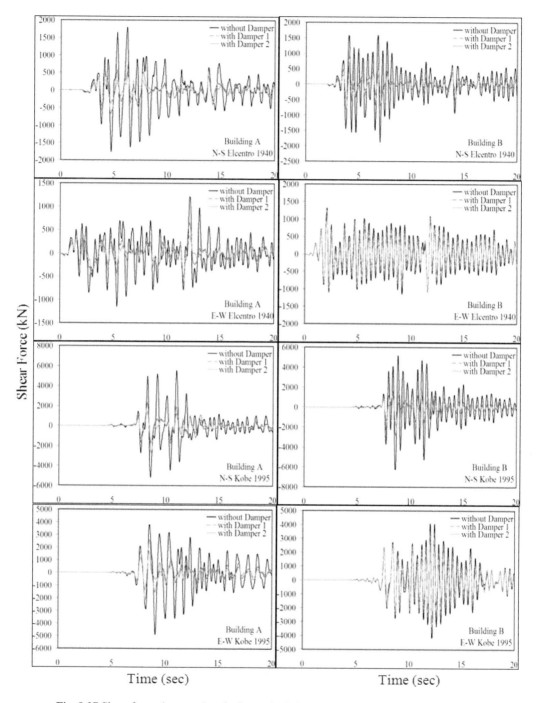

Fig. 8.27 Shear force-time graphs of adjacent buildings for Example 2(b) in two directions.

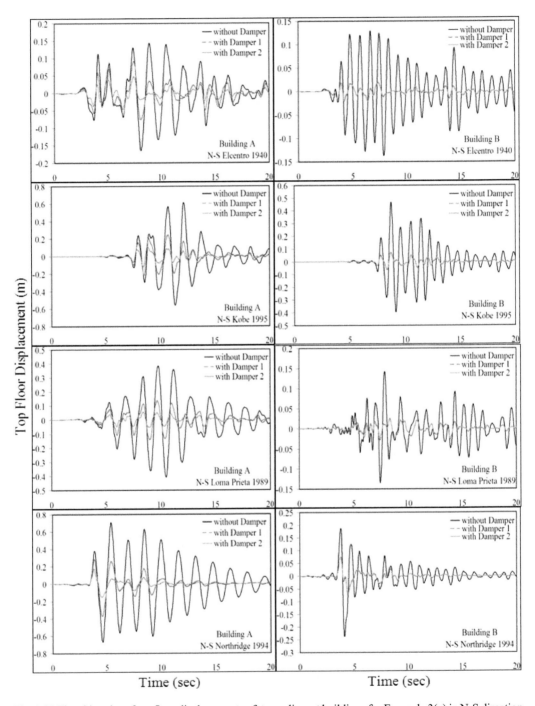

Fig. 8.28 Time histories of top floor displacements of two adjacent buildings for Example 3(a) in N-S direction.

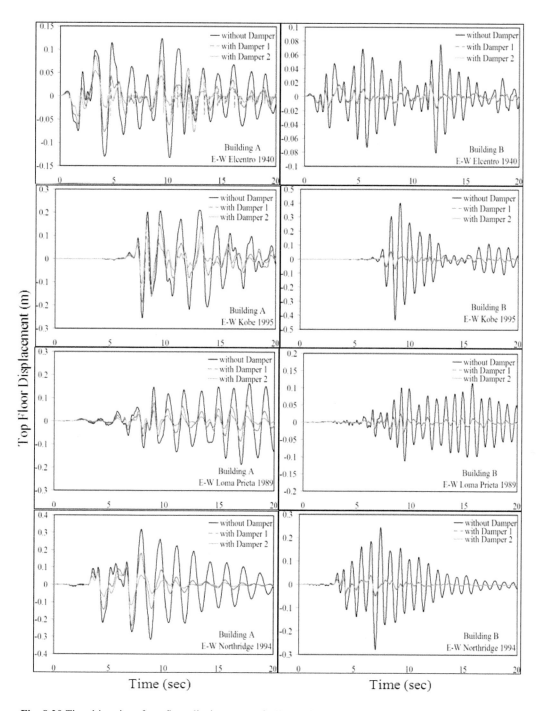

Fig. 8.29 Time histories of top floor displacement of adjacent buildings for Example 3(a) in E-W direction.

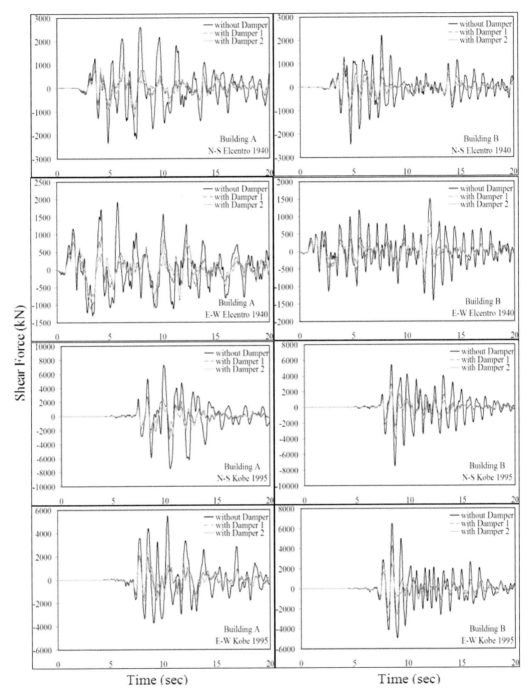

Fig. 8.30 Shear force-time graphs in N-S and E-W directions for Example 3(a).

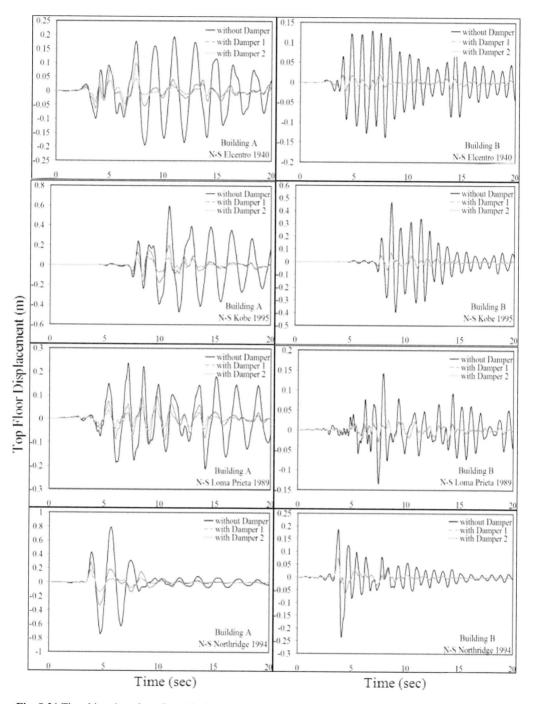

Fig. 8.31 Time histories of top floor displacement of adjacent buildings for Example 3(b) in N-S direction.

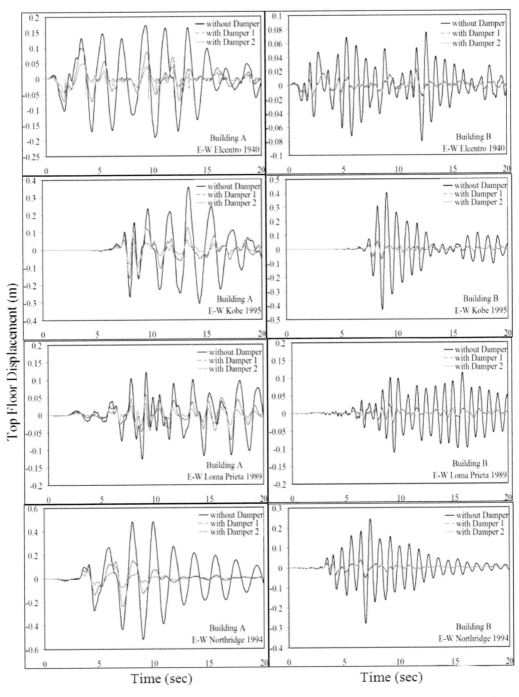

Fig. 8.32 Time histories of top floor displacement of adjacent buildings for Example 3(b) in E-W direction.

It can be noted that top floor displacements for Building *A* do not change significantly in E-W Loma Prieta 1989. In Fig. 8.33, shear forces of both buildings are not different even though the adjacent buildings are connected with different dampers in terms of damping coefficients.

For Example 4(a), Fig. 8.34 shows the time histories of the top floor displacement of adjacent buildings for all earthquakes in N-S direction, respectively, with and without the joint dampers. For adjacent buildings connected with Damper 1 or Damper 2, it is seen that there are no differences in the amplitudes of displacement for N-S direction as being different from Example 1(a).

Within the first nine or eleven seconds, the amplitudes of top floor displacement of Building *A* in N-S Elcentro 1940 reduce significantly, while the amplitudes in the same earthquake do not reduce. In addition, after the first eleven seconds, the peak responses of buildings in N-S direction in both the earthquakes reduce with the peak response reduction range being between 30–60 per cent.

Furthermore, Example 4 shows that the efficacy of dampers for coupled buildings with different shear stiffness but with same elevations is more than the adjacent buildings with the same stiffness and height.

Fluid viscous dampers can reduce significantly the amplitudes of displacement in N-S direction because of different shear forces of each building, especially 1994 Northridge earthquake. Figure 8.36 demonstrates the shear force-time graphs for Example 4(a) in N-S and E-W directions using Northridge 1994 and Loma Prieta 1989 earthquakes.

The dampers can mitigate the amplitudes of shear forces in both the earthquakes. After the first ten seconds, the peak responses of buildings in E-W direction in both the earthquakes reduce with the peak response reduction ranging between 10–20 per cent.

For Example 4(b), the results of time histories of top floor displacements and the base shear force responses of two 20-storey buildings are shown in Fig. 8.37 and Fig. 8.38 with all the earthquakes. It is shown in Fig. 5.37 that the amplitudes of displacement in N-S direction within 15 seconds are mitigated. However, the peaks reduce significantly with the peak response reduction ranging between 30–70 per cent, although the adjacent buildings have the same height. In E-W direction, there is a difference in terms of the response between the unlinked and linked Buildings *A* and *B*.

Figure 8.38 indicates the shear force-time graphs of adjacent buildings for Example 4(b) in two directions under Kobe 1995 and Elcentro 1940 earthquakes. The reduction of shear forces for Example 4(b) is more than that for Example 1(b). The amplitudes of forces for earthquakes mentioned above in E-W direction do not reduce.

It can be seen from all the examples discussed above that linked buildings by dampers on all floors are more effective than unlinked buildings for mitigation of earthquake effects. The following section investigates the optimum placement of dampers, instead of placing them on all the floors.

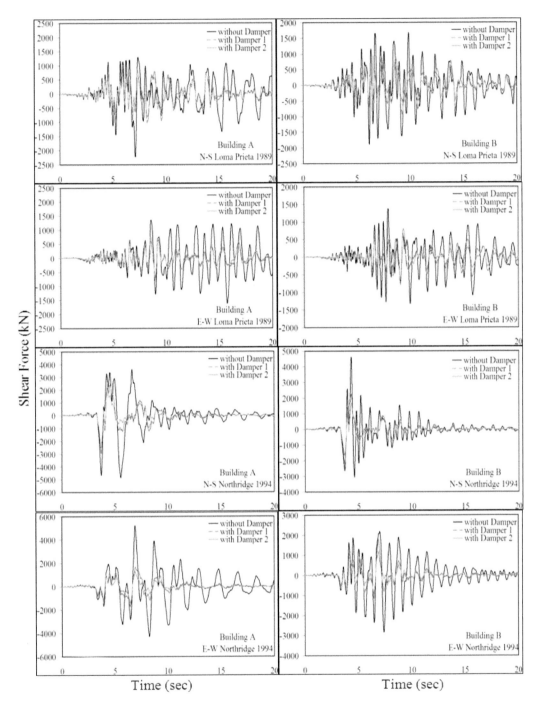

Fig. 8.33 Shear force-time graphs of adjacent buildings for Example 3(b) in two directions.

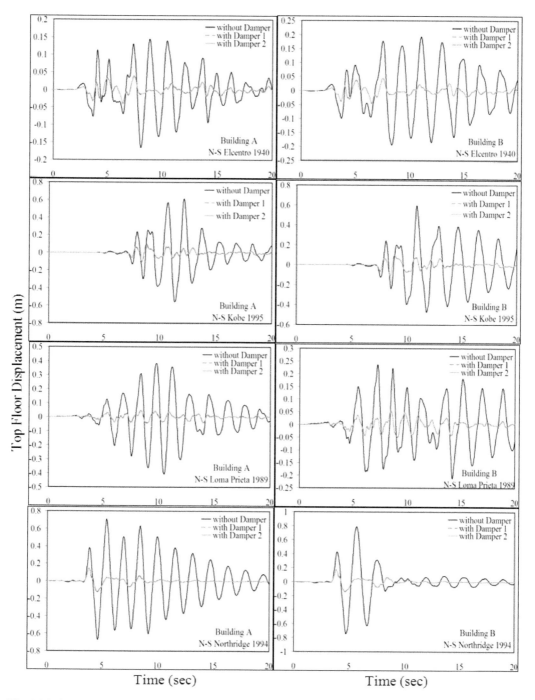

Fig. 8.34 Time histories of top floor displacements of two adjacent buildings for Example 4(a) in N-S direction.

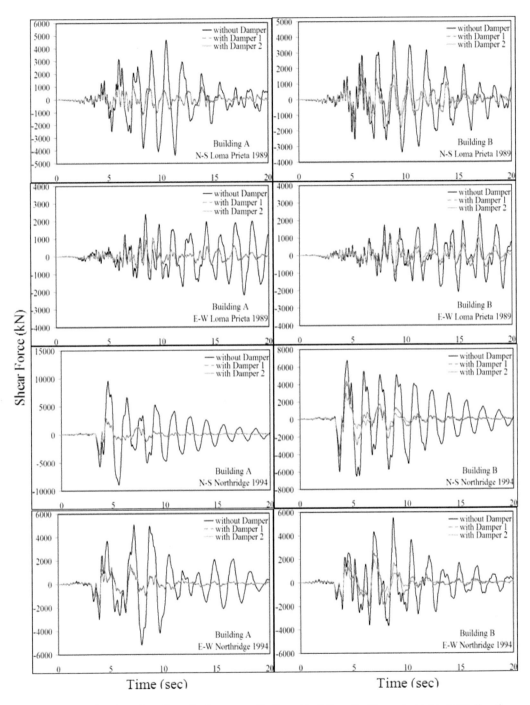

Fig. 8.35 Time histories of top floor displacement of adjacent buildings for Example 4(a) in E-W direction.

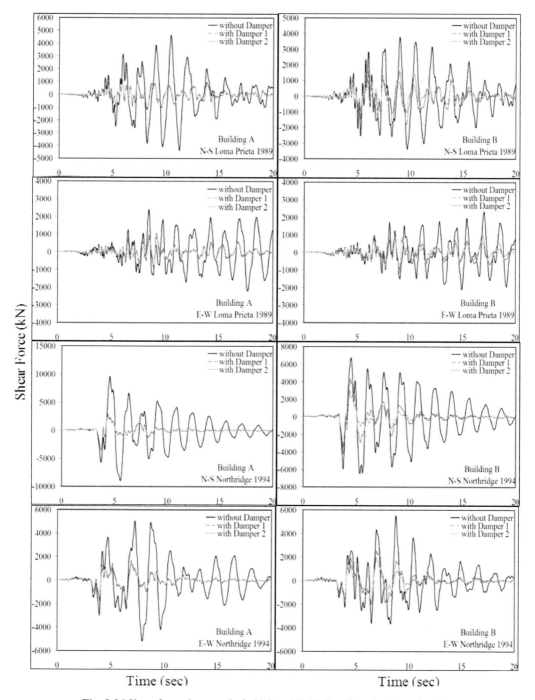

Fig. 8.36 Shear force-time graphs in N-S and E-W directions for Example 4(a).

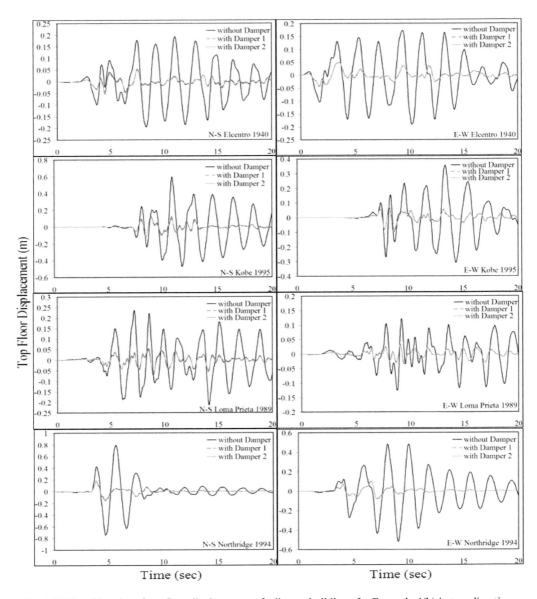

Fig. 8.37 Time histories of top floor displacement of adjacent buildings for Example 4(b) in two directions.

8.4 Summary

In this chapter, the results of all examples are evaluated, based on the reduction of displacement, acceleration and shear force responses of adjacent buildings. The numerical results are carried out in two groups, namely, frequency domain and time domain.

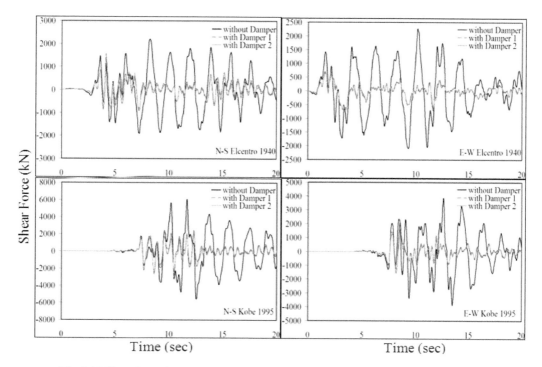

Fig. 8.38 Shear force-time graphs of adjacent buildings for Example 4(b) in two directions.

Firstly, frequency domain is graphically evaluated in terms of the spectral density functions of displacement and acceleration for the four example buildings. In Example 2 and Example 3, it is clearly seen that by using the damper for the lower Building *B* is more beneficial than that for the higher Building *A*. There is a trend that the joint dampers are more useful for lower adjacent buildings than for higher adjacent buildings. In Example 2(b) and Example 3(b), it is observed that the peaks have slowly change, although the buildings have the same dynamic characteristics.

Secondly, time domain is graphically evaluated in terms of the time histories of base shear force and displacement for the four buildings. In Example 1 and Example 4, the amplitudes of displacement reduce significantly in N-S direction, although the unlinked and linked buildings have different reduction in the amplitude of displacement in E-W direction. The maximum reduction of top floor displacement is 50 per cent in Example 2, while the reduction is almost 35 per cent in Example 3. Example 4(b) shows that use of damper for high adjacent buildings with same characteristics cannot reduce to the amplitude of displacements. Example 4(a) shows that absolute displacements in terms of floor number are mitigated by using fluid dampers for high adjacent buildings with different shear stiffness.

9

Results for Optimum Placement of Dampers

9.1 Introduction

In order to minimise the cost of dampers, the responses of two adjacent buildings are investigated by considering only three dampers (almost 50 per cent of the total) with optimum damper properties obtained by Xu et al. (1999) at selected floor locations. For locations of dampers, the floors with maximum relative displacement are selected. Many trials are carried out to arrive at the optimal placement of dampers. The graphs shown below are the variations of the displacements in all the floors for different cases. It is seen that the maximum displacement values in original duration of 60s taken at a total of 3,000 time records at an interval of 0.02s are selected for the graphs below. To illustrate the overall effectiveness of fluid viscous dampers on adjacent buildings, the standard deviations of displacement at each floor for each building with and without dampers are indicated by using selected earthquakes as below.

9.2 Results of Optimum Placement of Dampers

Figure 9.1 shows the four cases which were investigated for Example 2; for the remaining examples, Case (i) represents the control case where the buildings are not connected. In Case (ii), the dampers are placed in all floors. The dampers in Case (iii) are placed at odd floors. Finally, the dampers in Case (iv) are placed in the floors above the middle of the shorter buildings.

For Example 1(a), Fig. 9.2 shows the variation of absolute displacements, namely, when Case (i) unconnected, Case (ii) connected at all the floors, Case (iii) connected at Floors 3, 4 and 5 and Case (iv) connected at Floors 1, 3 and 5. It is observed from the figures that

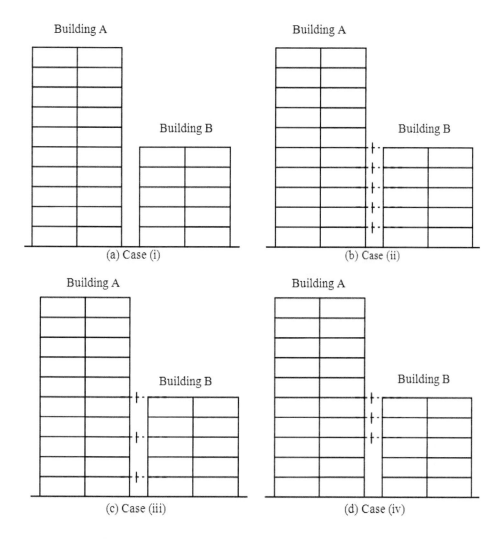

Fig. 9.1 Locations of dampers in adjacent buildings of Example 2.

the dampers are more effective when they are placed at Floors 3, 4 and 5. For the purpose of occurring cases, the time history of relative horizontal displacements at the top level of two 5-storey buildings are indicated for uncontrolled and controlled adjacent buildings in Fig. 9.2. When the dampers are located on these floors, the displacements in all the storeys are reduced almost as much as when they are attached at all the floors. Hence, Floors 1, 3 and 5 are considered to be the optimal placement for dampers. It shows that the dampers at appropriate placements can alleviate considerably the seismic responses of the coupled system, besides reducing the cost of the dampers to a greater level.

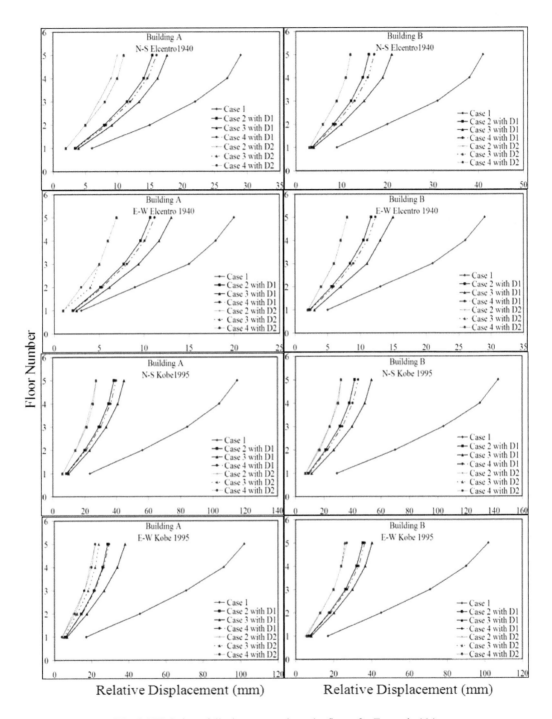

Fig. 9.2 Variation of displacements along the floors for Example 1(a).

174

For Example 1(b), use of damper in adjacent buildings with the same heights is more beneficial when Case 4 with Damper 2 is provided for adjacent buildings, as shown in Fig. 9.3. However, Case 3 with Damper 2 can be more beneficial under some earthquakes such as Loma earthquake and Northridge earthquake in E-W direction.

For Example 2(a), the variation of absolute displacements are indicated in Fig. 9.4, namely, when Case (i) unconnected, Case (ii) connected at all the floors, Case (iii) connected at Floors 6, 7, 8, 9 and 10 and Case (iv) connected at Floors 1, 3, 5, 7, 9. It is interesting that the dampers can be more effective when placed on Floors 3, 4 and 5 for Building *B*. However, for Building *A*, the use of damper is not more effective when placed at selected floors.

Moreover, the amplitudes of displacement increase for Building *A* in Kobe 1995 and Elcentro 1940, use of Case 3 with Damper 1. It is observed that Case (ii) is more suitable for this example among all the cases. As shown in Fig. 9.5, use of damper at Case (iv) with Damper 2 in N-S Loma Prieta 1989 and N-S Northridge 1994 for both buildings can be more suitable in reduction of displacement than in other cases.

For Example 3(a), the variation of absolute displacements is indicated in Fig. 9.6, namely, when Case (i) unconnected, Case (ii) connected at all the floors, Case (iii) connected at Floors 10, 11, 12, 13, 14, 15, 16, 17, 18, 19 and 20 and Case (iv) connected at Floors 1, 3, 5, 7, 9, 11, 13, 15, 17 and 19. It is also interesting that the dampers for adjacent buildings with different elevations can be more effective when placed at selected floors (Case iv) in N-S Elcentro 1940 and N-S Kobe 1995, especially in low-rise buildings.

However, for earthquakes in E-W direction, the use of damper cannot be more effective when placed at any floors for Kobe 1995. The amplitudes of displacement increase when using the damper for coupled buildings. For buildings having the same dynamic characteristics but different heights, the fluid viscous dampers cannot mitigate earthquake effects during some big earthquakes, as shown in Fig. 9.7.

Use of dampers at selected floors for Example 2 is more effective than Example 3. For Example 4(a), the variation in absolute displacements in terms of floor numbers is demonstrated in Fig. 9.8. All cases are the same as Example 3. It is also interesting that the dampers for adjacent buildings with different shear stiffness can be more effective when placed at all floors, expect in the N-S Kobe 1940 earthquake. However, for a related earthquake, the use of damper is more effective when placed at selected floors except in Case 1.

Moreover, the amplitudes of displacement decrease for coupled buildings via the damper located at all floors. Figure 9.9 indicates variation of displacements along the floors for Example 4(b) in two directions. Example 4(a) and 4(b) show that buildings with different shear stiffness are more effective than buildings having the same shear stiffness on reduction of earthquake effects, using fluid viscous dampers in Cases ii, iii and iv.

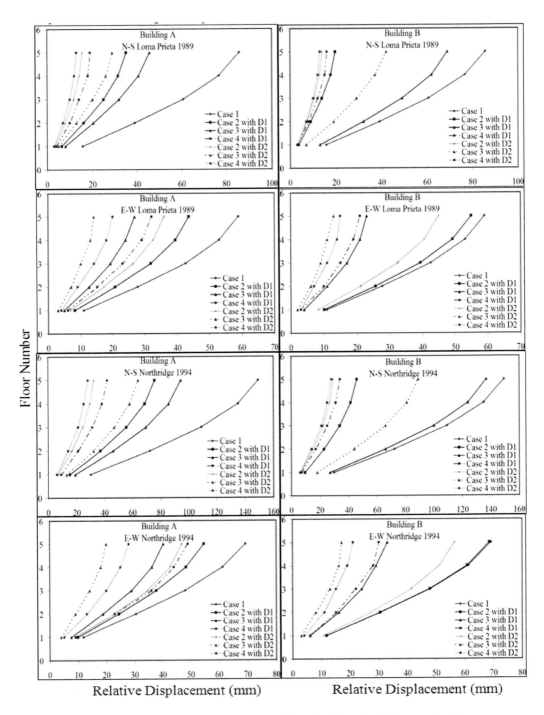

Fig. 9.3 Variation of displacements along the floors for Example 1(b) in two directions.

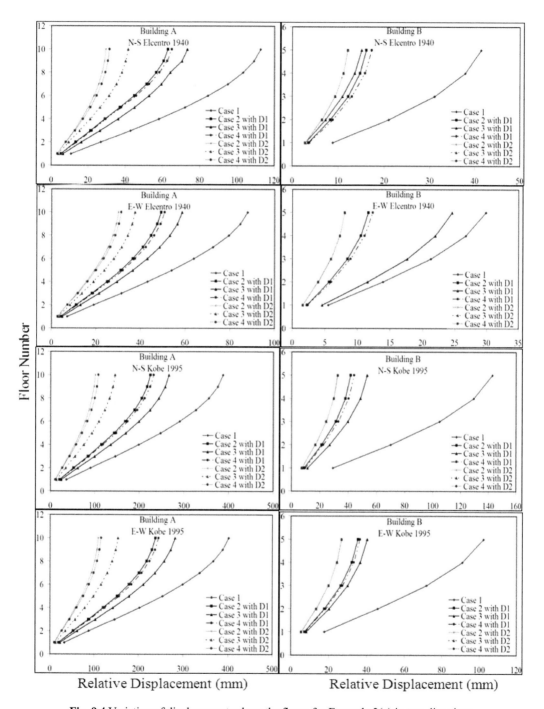

Fig. 9.4 Variation of displacements along the floors for Example 2(a) in two directions.

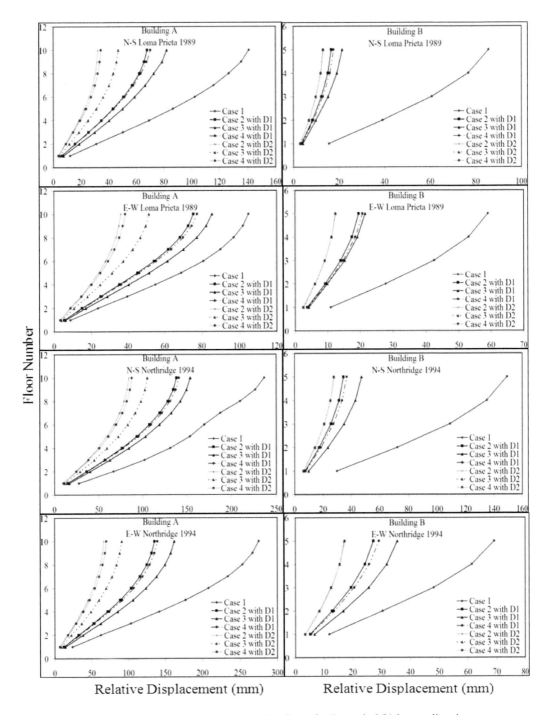

Fig. 9.5 Variation of displacement along the floors for Example 2(b) in two directions.

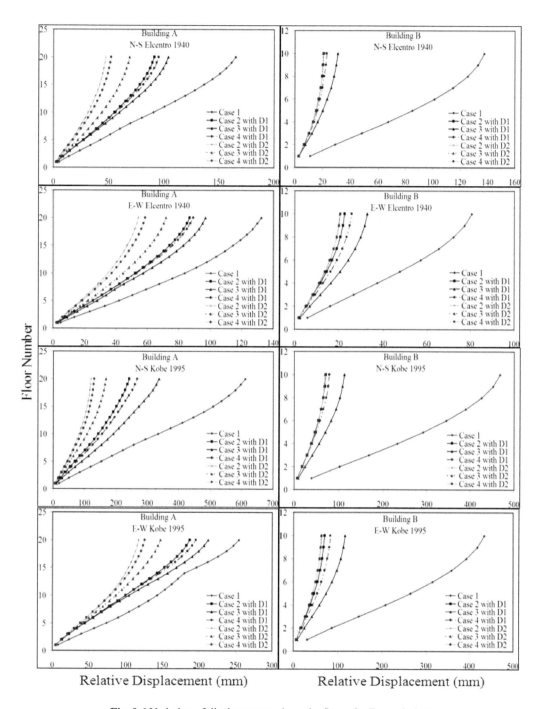

Fig. 9.6 Variation of displacements along the floors for Example 3(a).

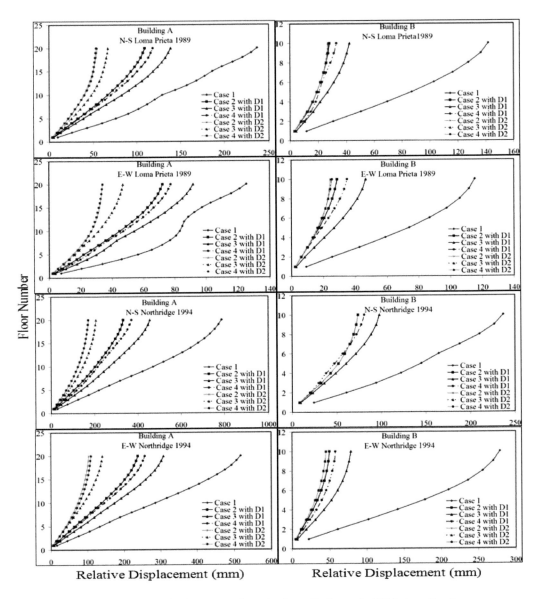

Fig. 9.7 Variation of displacements along the floors for Example 3(b) in two directions.

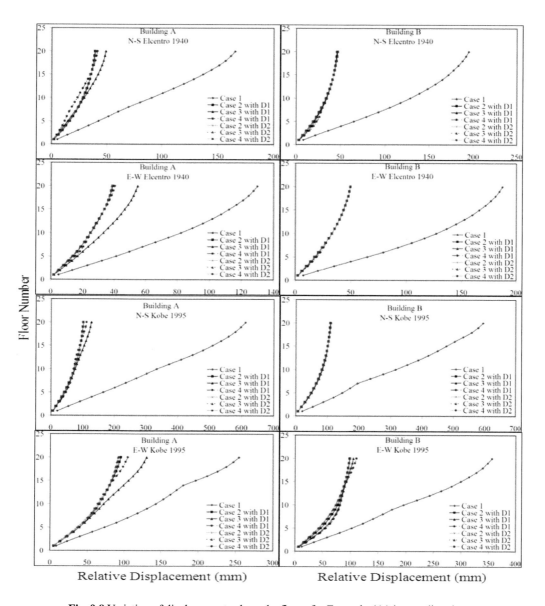

Fig. 9.8 Variation of displacements along the floors for Example 4(a) in two directions.

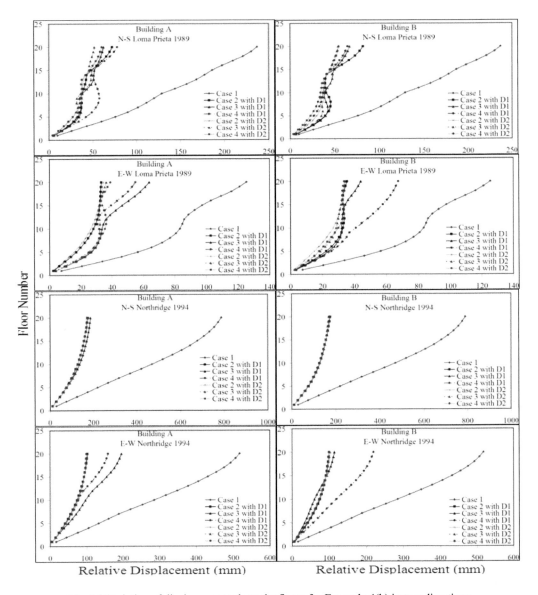

Fig. 9.9 Variation of displacements along the floors for Example 4(b) in two directions.

9.3 Summary

The variation in absolute displacements for the cases mentioned above is shown to favour optimum placement of dampers. All examples calculated in each case show that the dampers at suitable placements can reduce significantly the seismic responses of the coupled system, besides reducing the cost of dampers to a greater level.

10
Optimum Design Examples

10.1 Introduction

In this chapter, six numerical examples are described. In Section 10.2, the numerical model for 4 and 3 storey base-isolated coupled buildings is used to investigate the effects of pounding, which is a modification of the examples adopted by Jankowski (2008). In Section 10.3, the 5 and 4 storeys fixed-base asymmetric buildings are used in order to investigate the SSI effects and pounding through the rigorous method. Two numerical examples are designed for adjacent buildings, utilising passive and active dampers in Section 10.4. For utilising the capabilities of MR dampers on to adjacent buildings, two numerical examples are presented in Section 10.5. The responses of these numerical examples are investigated under strong components of various earthquake excitations based on peak ground accelerations. Details of earthquake excitations used in this study are also given in this chapter.

10.2 Base-isolated Buildings

The dynamic equations in Eq. 2.1 for the validation of numerical models can be conducted to analyse substantially different dynamic properties of adjacent building systems. Based on the fourth order Runge-Kutta method, a MATLAB program is developed to solve the equations of motion for the system subjected to excitations of earthquake ground acceleration and numerical simulations presented are used for a parametric study.

10.2.1 Description of Model

The following basic values describing the structural characteristics adopted by Jankowski (2008) in Table 10.1 are used for the storey yield strengths, F_{xi}^y, F_{yi}^y and mass of the base of two buildings:

For Building A:

$$F_{x1}^y = F_{x2}^y = F_{x3}^y = F_{x4}^y = F_{y1}^y = F_{y2}^y = F_{y3}^y = F_{y4}^y = 1.369 \times 10^5 \, \text{N},$$

$$m_{B1} = 37.134 \times 10^3 \, \text{kg},$$

For Building B:

$$F_{x5}^y = F_{x6}^y = F_{x7}^y = 1.442 \times 10^7 \, \text{N},$$

$$F_{y5}^y = F_{y6}^y = F_{y7}^y = 1.589 \times 10^7 \, \text{N},$$

$$m_{B2} = 15.246 \times 10^5 \, \text{kg}.$$

As shown in Table 10.2, the first natural vibration periods of both buildings in the longitudinal, transverse, and vertical direction are shown respectively. The damping matrix can be obtained by assuming that it is proportional to the stiffness matrix (Clough and Penzien 1993). The dashpot constants for the building can be written and simplified as shown in Eqs. 2.16a–d and 2.17a–b. By assuming 5 per cent damping in the first mode, a damping ratio in the second mode can be found in Eq. 2.17a–b for the reference building.

Table 10.1 Structural Characteristics of Base Isolated Buildings.

Building A (Reference Building)						
Storey No.	Mass, m (kg) (10^3)	Stiffness, k (N/m)			Damping Coefficient, c (kg/sec)		
		x (10^6)	y (10^6)	z (10^{10})	x (10^4)	y (10^4)	z (10^6)
1	25	3.46	3.46	1.246	6.609	6.609	3.969
2	25	3.46	3.46	1.246	6.609	6.609	3.969
3	25	3.46	3.46	1.246	6.609	6.609	3.969
4	25	3.46	3.46	1.246	6.609	6.609	3.969
Building B (Heavier and Stiffer)						
Storey No.	m (kg) (10^6)	k (N/m)			c (kg/sec)		
		x (10^9)	y (10^8)	z (10^{11})	x (10^7)	y (10^6)	z (10^8)
1	1.0	2.215	5.537	2.215	1.058	5.286	1.058
2	1.0	2.215	5.537	2.215	1.058	5.286	1.058
3	1.0	2.215	5.537	2.215	1.058	5.286	1.058

Table 10.2 Properties of Buildings in the Longitudinal, Transverse and Vertical Directions.

Properties	Building A			Building B		
	x	y	z	x	y	z
First mode time period (sec)	1.54	1.54	0.026	0.3	0.6	0.03
Second mode time period (sec)	0.53	0.53	0.009	0.11	0.21	0.01
First mode frequency (mod/sec)	4.08	4.08	245.2	20.94	10.47	209.44
Second mode frequency (mod/sec)	11.76	11.76	706	59.2	29.3	5886.9

In order to enhance the accuracy of the analysis, the coefficient of restitution can be determined separately for each collision, depending on the relative prior-impact velocity of structures (Robert 2009). However, the constant value of coefficient of restitution is used in numerical models during the entire time of the ground motion so as to obtain the general pounding-involved structural response. According to the results obtained by Jankowski (2006), $\bar{\beta} = 2.75 \times 10^9$ N/m$^{3/2}$, $\bar{\xi} = 0.35$ in Eq. 2.10a–d were applied for values of non-linear visco-elastic pounding force model's parameters. The coefficient of restitution, which accounts for energy dissipation during collision, is expressed for concrete to concrete impact as:

$$e = -0.007v^3 + 0.0696v^2 - 0.2529v + 0.7929 \tag{10.1}$$

where v is the prior-impact relative velocity of two colliding bodies. In fact, the coefficient of restitution, e, ranging from 0.5 to 0.75 provides reasonable engineering approximation for studying structural response with pounding (Robert 2009). Hence, the value of e = 0.65 has been arbitrarily chosen. The coefficient of friction of the sliding bearing remains constant throughout the motion of the structure, even though the coefficient of friction is dependent on the pressure and sliding velocity (Bhasker Rao and Jangid 2001). The value of the friction coefficient can be calculated by Eq. 10.2 which was obtained by Constantinou et al. (1990).

$$\mu u_i = f_{max} - \Delta f \times e^{-a|\dot{U}|} \tag{10.2}$$

where f_{max}, Δf, a, and $\dot{U}$ are the coefficient of friction at large sliding velocity, the differences between f_{max} and the coefficient of friction at low sliding velocity, the constant value and the sliding velocity, respectively. The value of friction coefficient, μ_f, has been used as 0.5, whereas the value of friction coefficient of the sliding bearing, μu_i, has been used as 0.10 (Wriggers 2006b). The initial gap, D, between the buildings is taken as 0.02 m. The El-Centro (18.05.1940) and the Duzce (12.11.1999) earthquake records are recorded as the input with the N-S, E-W, and U-D components of the ground motion in the longitudinal, transverse and vertical directions as shown in Table 10.3, respectively.

When the contact of the buildings in the longitudinal direction is detected, the pounding forces in the transverse and vertical directions are applied.

Table 10.3 Earthquake Records Used in this Study.

Earthquakes	MW	Station	PGA (g) (N-S, E-W, U-D)	Duration (sec)
El Centro, U.S.A. 18/05/1940	7.0	117 El Centro Array-9	0.313, 0.215, 0.205	39.99
Duzce, Turkey 12/11/1999	7.1	375 Lamont	0.97, 0.514, 0.193	41.50

10.3 Fixed Adjacent Buildings

In order to investigate the influence of modelling, the structural behaviour by using either elastic or inelastic systems on the response of asymmetric adjacent buildings considering the SSI system resting on the soft and hard soils, the dynamic equations derived in the most general form in Eq. 2.15a–c for validation of the numerical models is conducted under different ground excitations. The excitations of El-Centro 1940 earthquake record in Table 10.3 are examined for the seismic response of coupled buildings. Both asymmetric Building *A* and Building *B* resting on an elastic half-space are considered as five and four storey buildings, respectively. The details of the SSI system are briefly presented herein.

10.3.1 Model Description

The dimensions of both Building *A* and Building *B* are rectangular in plan with 20 m × 15 m and 25 m × 20 m, the larger plan dimensions being parallel to the longitudinal direction (x) for each building, respectively. The ratio of the base mass to the floor mass of the buildings is 3 for each building. Moreover, for translation in the *x* and *y* axes, twist about the *z* axis and rocking about the *x* and *y* axes, the dimensions of rectangular base of adjacent buildings can be converted into an equivalent circular base (r_o) having the same area as the plan of each building based on the formulas determined by Richart et al. (1970). Hence, the calculations of the radius of base mass determined by Richart et al. (1970) are used, considering translations, rotation and rocking directions, herein. The height of each storey is 2.85 m in both the buildings. According to the results obtained by Jankowski (2006, 2008, 2010), $\bar{\beta} = 2.75 \times 10^9$ N/m$^{3/2}$ and $\bar{\xi} = 0.35$ were applied for values of non-linear visco-elastic pounding force model's parameters. The moment of inertia of the rigid body for each building about the centroidal axes parallel to the *x* and *y* axes are evaluated by replacing each floor with a disc of radius (r_o). The following basic values describing the structural characteristics in Table 10.4 have been used:

Table 10.4 Structural Characteristics of Buildings.

Storey No.	Height of Floor level $h_{i,j}$ (m)	Building *A*		Building *B*	
		$m_i \times 10^6$ (kg)	$k_i \times 10^8$ (N/m)	$m_i \times 10^6$ (m)	$k_i \times 10^8$ (N/m)
1F	2.85	0.30	3.46	0.4065	5.06
2F	5.7	0.30	3.46	0.4065	3.86
3F	8.55	0.30	3.46	0.4065	3.86
4F	11.4	0.30	3.46	0.4065	3.86
5F	14.25	0.30	3.46	–	–

The translational stiffness in transverse direction and the torsional stiffness around the centre of mass for each storey of each building are proportional to the stiffness in the longitudinal direction of the same storey and given by the following formula in Eq. 10.3a–d (Kan and Chopra 1976).

$$\beta_y = \frac{k_{yi}}{k_{xi}}, \beta_y = \frac{k_{yj}}{k_{xj}}, \beta_t = \frac{k_{\theta i}}{r_a^2 k_{xi}}, \beta_t = \frac{k_{\theta j}}{r_b^2 k_{xj}} \qquad (10.3a–d)$$

The ratios, β_y and β_t are taken as 1.32 and 1.69 for both the buildings, respectively. The constant of proportionality, α in Eq. 2.16a–d is evaluated on the basis of 2 per cent of critical damping in the fundamental mode of superstructures in both the buildings.

10.3.2 Properties of the SSI Systems

The density of soil medium, ρ and Poisson's ratio, v are taken to be 1922 kg/m³ and 0.333, respectively. In order to examine the effectiveness of the rigorous method for the entire SSI systems, using the direct integration method to solve the equations of motion, four soil types are investigated in the range of shear velocities, v_s of 65 m/sec (soft soil), 130, 200 and 300 m/sec (hard soil) have been specifically chosen for this study as Case I to IV, respectively. Case I (v_s = 65 m/sec) and Case IV (v_s = 300 m/sec) are created for the SSI systems resting on soft and hard soils in order to investigate the seismic response of adjacent buildings under large and small SSI effects, respectively. With reference to Table 2.1, the values of stiffness and damping coefficients in the translation in the x and y axes, twist about the z axis and rocking about the x and y axes for each case in both the Building A and Building B are listed in Table 10.5.

These cases are subjected to NS and EW components of the 1940 El-Centro earthquake record along the x and y axes, respectively. Figure 10.1 presents the influence of chosen cases onto the SSI systems of coupled buildings under excitation of the 1940 El-Centro earthquake. Based on a dimensionless frequency, a_0 resulting from $0 \le a_0 = \omega_f r_0/v_s \le 1.5$, the maximum wave frequencies of both are f_{max} = 2.1 Hz, 9.7 Hz of Building A, 1.8 Hz and 7.8 Hz of Building B for Case I and Case IV, respectively. As can be seen in Fig. 10.1, most of the energy of related ground motion is at frequencies less than 1.8 Hz. Hence, by using frequency independent spring and dashpot set for coupled buildings, the SSI effects of the chosen cases can be conducted effectively. Furthermore, each coupled building is modelled as reference building resting on a rigid base with a similar superstructure. Table 10.6 shows the properties of the reference buildings in all the directions. It can help to see how the SSI affects the dynamic characteristics of coupled buildings. The initial gap, D, between the buildings is taken as 0.04 m.

Table 10.5 Impedance Values for Case I and Case IV in Both the Buildings.

Stiffness Coefficients	Building A		Building B		Damping Coefficients	Building A		Building B	
	I	IV	I	IV		I	IV	I	IV
K_T 10^7 (kN/m)	0.04	0.83	0.05	1.07	C_T 10^5 (kN sec/m)	0.34	1.56	0.56	2.60
K_θ 10^9 (kNm)	0.04	0.92	0.09	1.95	C_θ 10^7 (kNmsec)	0.10	0.46	0.27	1.26
K_ψ 10^9 (kNm)	0.03	0.54	0.06	1.22	C_ψ 10^7 (kNmsec)	0.09	0.42	0.30	1.38
K_ϕ 10^9 (kNm)	0.04	0.83	0.08	1.70	C_ϕ 10^7 (kNmsec)	0.17	0.79	0.47	2.18

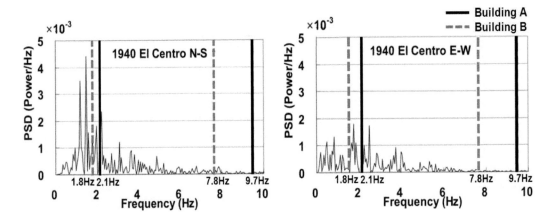

Fig. 10.1 Power spectral densities for the NS-EW components of the 1940 El-Centro earthquake.

Table 10.6 Properties of Buildings in the Longitudinal, Transverse and Vertical Directions.

Properties	Building A			Building B		
	x	y	z	x	y	z
First mode time period (sec)	0.65	0.57	0.07	0.56	0.5	0.05
Second mode time period (sec)	0.22	0.19	0.023	0.19	0.21	0.02
First mode frequency (mod/sec)	9.67	11.11	90.69	11.29	12.97	135.60
Second mode frequency (mod/sec)	28.22	32.42	264.7	32.22	37.02	387.13

In order to consider the effect of pounding on the coupled buildings, when the contact of the buildings in the longitudinal direction is detected, the pounding forces in the transverse and vertical directions are applied.

For the response time histories of the entire SSI system of coupled buildings, the rigorous method which uses the direct integration method to solve the equation of motion as shown in Eq. 2.15a–c is called Rig.

10.4 Numerical Study of Controlled Adjacent Buildings with Passive and Active Control Systems

This study investigates the efficacy of optimal passive and active dampers for achieving the best results in seismic response mitigation of adjacent buildings connected to each other by either passive or active dampers. The adjacent buildings in this numerical study are subjected to the 1940 El Centro (117 El-Centro Array-9 station) and the 1995 Kobe (KJMA station) excitations where the maximum ground acceleration scaled to 0.3 g for 1940 El Centro NS and 0.8 g for 1995 Kobe NS, as shown in Fig. 10.2.

The optimisation to minimise the H_2 and H_∞ norms in the performance indices are carried out by genetic algorithms (GAs). The effect of the uncontrolled case, passive and active control techniques are investigated on the behaviour of adjacent structures for controlling vibration. The first example has adjacent buildings connected by passive damper systems; the second is adjacent buildings utilising active control systems. The first example utilises fluid viscous damper to show the effect of optimum damper parameters in reduction of responses of adjacent buildings. Furthermore, in order to enhance the capability of

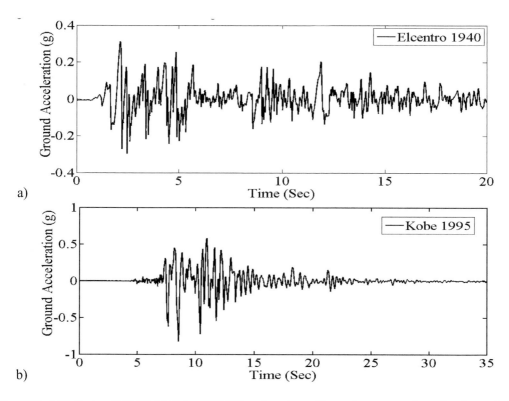

Fig. 10.2 a) El Centro 1940 NS b) Kobe 1995 NS earthquakes with maximum ground acceleration scaled between 0.3 g and 0.8 g, respectively.

Table 10.7 Genetic Algorithm Parameters Used in this Study.

GA Parameters	Binary Coding	Real Coding
Number of generations	250 or 1000	1000
Population	30	20
Probability of crossover	0.5	0.8
Probability of mutation	0.01	0.01
Number of new random chromosomes to be inserted after crossover and mutation (%)	20	10

passive damper, an actuator is installed at the top floor level between the buildings in the second example. In this study, numerical examples' parameters used by Arfiadi (2000) are slightly modified before adoption. Furthermore, for a passive control system, both binary and real coding are used and compared, to optimise the passive device parameters. For active control system, real coded GA is used by defining the regulated output to obtain the controller gains. Hence, the controller gains based on the available measurement outputs are obtained. By using real coded GA in H_∞ norm, the optimal controller gain is obtained by different combinations of measurements as feedback for designing control force between adjacent buildings. The GA parameters used in this study are separated into two sections, as shown in Table 10.7.

The optimisation problem means finding the optimum of damper parameters that minimise the response of buildings. The procedure of GA is used in this study as an efficient tool.

10.4.1 GA-H_2 and GA-H_∞ Optimisation Procedures

For optimisation of passive and active control problems between adjacent buildings, several optimisation methods based on the chosen objective function have been synthesised in this study. The H_2 and H_∞ norms are considered as objective functions in the modern control theory to minimise the transfer function from external disturbance to regulated output.

$$f = \alpha \times (1/J) \tag{10.4}$$

where J is the objective function and α is a constant value to scale the fitness function and is taken as 10 in this numerical study. The conversion of objective function in the form of Eq. 10.4 is possible on the assumption that H_2 and H_∞ norm transfer functions are positively definite. The first objective of Example 1 is to determine the optimum value of stiffness k_d and the damping coefficient c_d of the dampers according to the performance index and regulated output. After forming the equation of motion, state equation in terms of state vector X is shown in here. For H_2 norm optimisation, in order to obtain passive control parameters (c_d, k_d),

$$\dot{X} = AX + E\ddot{X}_g(t)$$

$$x = C_w X$$

$$J = \left[tr\left(C_w L_c C_w^T\right)\right]^{1/2} = \left[tr\left(E^T L_o E\right)\right]^{1/2} \Rightarrow \text{Min.} \qquad (10.5\text{a–e})$$

$$AL_c + L_c A^T + EE^T = 0 \quad \text{or}$$

$$A^T L_o + L_o A + C_w^T C_w = 0$$

For H_∞ norm optimisation, in order to obtain passive control parameters (c_d, k_d),

$$\dot{X} = AX + E\ddot{X}_g(t)$$

$$x = C_w X$$

$$J = \max(\gamma) \Rightarrow \text{Min.} \qquad (10.6\text{a–d})$$

$$H = \begin{bmatrix} A + ER^{-1}D^T C_w & ER^{-1}E^T \\ -C_w^T\left(I + DR^{-1}D^T\right)C_w & -\left(A + ER^{-1}D^T C_w\right)^T \end{bmatrix}$$

H_2 or H_∞ norm transfer function is minimised with respect to different combinations of regulated outputs (C_w). In order to obtain the gain matrices G_a, G_v, G_d, the following procedure is conducted on the objective function to be used. For H_∞ optimal feedback control,

$$\dot{X} = A_{cl}X + E\ddot{X}_g$$

$$U(t) = -G_z X$$

$$x = C_w X$$

$$J = \max(\gamma) \Rightarrow \text{Min.} \qquad (10.7\text{a–e})$$

$$H = \begin{bmatrix} A_{cl} & ER^{-1}E^T \\ -C_w^T C_w & -A_{cl}^T \end{bmatrix}$$

$$A_{cl} = A - BG_z$$

Depending on both performance indices and regulated outputs, the optimal control gains can be obtained using GA. The design of a feedback control system has the flexibility for wider design constraints. When the design variables are discrete, the genetic algorithm selects the values required to be given, as described in the following Section 11.5.2.

10.4.2 Description of Model 1

In Example 1, a system of buildings located adjacent to each other and interconnected by passive dampers is considered to possess the optimal passive damper parameters.

Building *A* is a 6-storey shear building discussed in Arfiadi (2000). A 6-storey building discussed in Sadek et al. (1997) is taken as Building *B* (Therefore, N = 6 and M = 6 in Fig. 10.3). Note that same damper parameters are used at each floor level. In other words, the optimum passive damper parameters c_d, k_d obtained by using H_2 or H_∞ norm transfer functions in GA are used for each storey level. The structural parameters having mass, stiffness and damping coefficient are shown for both buildings in Table 10.8. In this research study, several controllers, as shown in Table 10.9, are designed by choosing different combinations of measurements as feedback. As can be seen from Table 10.9,

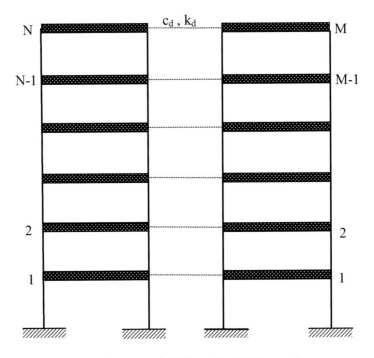

Fig. 10.3 N and M storey buildings with fluid viscous dampers.

Table 10.8 Structural Parameters of Both Buildings in Numerical Examples.

Floor (i)	Building A			Building B		
	m_i (t)	$K_i \times 10^5$ (kN/m)	$c_i \times 10^3$ (kN sec/m)	m_i (t)	$K_i \times 10^4$ (kN/m)	$c_i \times 10^3$ (kN sec/m)
6	514	3.5	1.190	134	4.679	0.604
5	542	3.5	1.190	143	4.991	0.644
4	542	3.5	1.190	152	5.302	0.684
3	542	3.5	1.190	161	5.614	0.724
2	542	3.5	1.190	170	5.926	0.752
1	542	3.5	1.190	179	6.247	0.806

Table 10.9 Objectives with the Corresponding Regulated Outputs.

Cases	Regulated Outputs
Case A	$C_w = \begin{bmatrix} \text{eye}(12) & \text{zeros}(12,12) \end{bmatrix}$
Case B	$C_w = \begin{bmatrix} \begin{bmatrix} 0 & 0 & 0 & 0 & 0 & 1 & 0 & 0 & 0 & 0 & 0 & 0 \\ 0 & 0 & 0 & 0 & 0 & 0 & 0 & 0 & 0 & 0 & 0 & 1 \end{bmatrix} & \text{zeros}(2,12) \end{bmatrix}$
Case C	$C_w = \begin{bmatrix} \text{zeros}(2,12) & \begin{bmatrix} 0 & 0 & 0 & 0 & 0 & 1 & 0 & 0 & 0 & 0 & 0 & 0 \\ 0 & 0 & 0 & 0 & 0 & 0 & 0 & 0 & 0 & 0 & 0 & 1 \end{bmatrix} \end{bmatrix}$
Case D	$C_w = \begin{bmatrix} 0 & 0 & 0 & 0 & 0 & 1 & 0 & 0 & 0 & 0 & 0 & 0 \\ 0 & 0 & 0 & 0 & 0 & 0 & 0 & 0 & 0 & 0 & 0 & 1 \end{bmatrix} \times \begin{bmatrix} -M^{-1}K & -M^{-1}C \end{bmatrix}$
Case E	$C_w = \begin{bmatrix} \begin{bmatrix} 1 & 0 & 0 & 0 & 0 & 0 & 0 & 0 & 0 & 0 & 0 & 0 \\ -1 & 1 & 0 & 0 & 0 & 0 & 0 & 0 & 0 & 0 & 0 & 0 \\ 0 & -1 & 1 & 0 & 0 & 0 & 0 & 0 & 0 & 0 & 0 & 0 \\ 0 & 0 & -1 & 1 & 0 & 0 & 0 & 0 & 0 & 0 & 0 & 0 \\ 0 & 0 & 0 & -1 & 1 & 0 & 0 & 0 & 0 & 0 & 0 & 0 \\ 0 & 0 & 0 & 0 & -1 & 1 & 0 & 0 & 0 & 0 & 0 & 0 \\ 0 & 0 & 0 & 0 & 0 & 0 & 1 & 0 & 0 & 0 & 0 & 0 \\ 0 & 0 & 0 & 0 & 0 & 0 & -1 & 1 & 0 & 0 & 0 & 0 \\ 0 & 0 & 0 & 0 & 0 & 0 & 0 & -1 & 1 & 0 & 0 & 0 \\ 0 & 0 & 0 & 0 & 0 & 0 & 0 & 0 & -1 & 1 & 0 & 0 \\ 0 & 0 & 0 & 0 & 0 & 0 & 0 & 0 & 0 & -1 & 1 & 0 \\ 0 & 0 & 0 & 0 & 0 & 0 & 0 & 0 & 0 & 0 & -1 & 1 \end{bmatrix} & \text{zeros}(12,12) \end{bmatrix}$
Case F	$C_w = \begin{bmatrix} \begin{bmatrix} 1 & 0 & 0 & 0 & 0 & 0 & 0 & 0 & 0 & 0 & 0 & 0 \\ -1 & 1 & 0 & 0 & 0 & 0 & 0 & 0 & 0 & 0 & 0 & 0 \\ 0 & -1 & 1 & 0 & 0 & 0 & 0 & 0 & 0 & 0 & 0 & 0 \\ 0 & 0 & -1 & 1 & 0 & 0 & 0 & 0 & 0 & 0 & 0 & 0 \\ 0 & 0 & 0 & -1 & 1 & 0 & 0 & 0 & 0 & 0 & 0 & 0 \\ 0 & 0 & 0 & 0 & -1 & 1 & 0 & 0 & 0 & 0 & 0 & 0 \\ 0 & 0 & 0 & 0 & 0 & 1 & 0 & 0 & 0 & 0 & 0 & 0 \\ 0 & 0 & 0 & 0 & 0 & 0 & 1 & 0 & 0 & 0 & 0 & 0 \\ 0 & 0 & 0 & 0 & 0 & 0 & -1 & 1 & 0 & 0 & 0 & 0 \\ 0 & 0 & 0 & 0 & 0 & 0 & 0 & -1 & 1 & 0 & 0 & 0 \\ 0 & 0 & 0 & 0 & 0 & 0 & 0 & 0 & -1 & 1 & 0 & 0 \\ 0 & 0 & 0 & 0 & 0 & 0 & 0 & 0 & 0 & -1 & 1 & 0 \\ 0 & 0 & 0 & 0 & 0 & 0 & 0 & 0 & 0 & 0 & -1 & 1 \\ 0 & 0 & 0 & 0 & 0 & 0 & 0 & 0 & 0 & 0 & 0 & 1 \end{bmatrix} & \text{zeros}(14,12) \end{bmatrix}$

193

several combinations of regulated outputs are used in this research study as follows: Case A: minimise displacements of both buildings, Case B: minimise displacements of the top floors of both buildings, Case C: minimise velocities of top floors of both buildings, Case D: minimise absolute accelerations of top floors of both the buildings, Case E: minimise inter-storey drifts of both buildings and Case F: minimise inter-storey drifts and displacements of top floors of both the buildings.

This study investigated not only the same damper properties used for each storey, but also the different dampers used at each floor level between the buildings. In this model, the adjacent buildings are only subjected to the 1940 El Centro (117 El-Centro Array-9 station) and the 1995 Kobe (KJMA station) excitations as shown in Fig. 10.2.

The damping matrix in Building A is assumed to be proportional to the stiffness matrix corresponding to about 1.5 per cent of the damping ratio of the first mode while Building B has 2 per cent of the damping ratio of the first mode. The performance index used in this model is as shown in Eq. 10.4 as GA tries to maximise the fitness function.

10.4.3 Description of Model 2

In numerical Example 2, the top floors of adjacent buildings are connected with only an active damper system, as shown in Fig. 10.4.

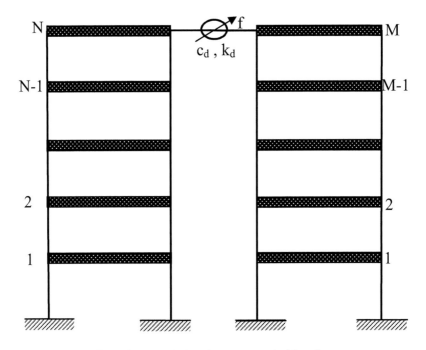

Fig. 10.4 Two adjacent buildings interconnected with active actuator.

The active control force is placed between adjacent buildings at the top floor level such that the damper now serves as a visco-elastic damper with an actuator. Further, the adjacent buildings are then subjected to El-Centro and Kobe excitations as in Example 1, but with the maximum ground acceleration scaled to 0.1 g and 0.3 g for both excitations to understand the capability of active systems. A visco-elastic damper is placed at the top floor-level between the buildings with $c_d = 515.63$ *kN sec/m* and $k_d = 3101$ *kN/m*. Note that the parametric values of visco-elastic damper are optimised by using the binary coding GA-H_∞ norm with Case F as the method in Example 1. In Example 2, the controller gains are obtained by using real coded GA with the performance index H_∞ norm. Details of several objectives with regulated outputs used in this study are given in Table 10.9. One actuator and four gains are used in Example 2. Hence, four design variables are to be determined. According to the chosen feedback in this research study, the gain matrix can be written for numerical Example 2 as referred in Eq. 3.20a–c.

$$G_d = \begin{bmatrix} 0 & 0 & 0 & 0 & 0 & G_{d1} & 0 & 0 & 0 & 0 & 0 & G_{d2} \end{bmatrix}$$

$$G_v = \begin{bmatrix} 0 & 0 & 0 & 0 & 0 & G_{v1} & 0 & 0 & 0 & 0 & 0 & 0 \end{bmatrix} \qquad (10.8a–c)$$

$$G_a = \begin{bmatrix} 0 & 0 & 0 & 0 & 0 & 0 & 0 & 0 & 0 & 0 & 0 & G_{a1} \end{bmatrix}$$

where G_{d1}, G_{d2}, G_{v1} and G_{a1} are gains to be determined in numerical Example 2. The closed loop system can be obtained as

$$\dot{X} = A_{cl}X + E\ddot{x}_g$$

$$A_{cl} = A - BG_z \qquad (10.9a–b)$$

The feedback proves the top floor displacement of both the buildings, the velocity of top floor of Building A and the absolute acceleration of the top floor of Building B. The regulated output C_w is in Case F and the control force can be written as

$$x = \begin{bmatrix} \alpha_1 \times C_w \\ -\alpha_2 \times G_z \end{bmatrix} X \qquad (10.10)$$

where α_1 and α_2 are constants to penalise the importance of each regulated output vector in Eq. 10.10. In this numerical study, $\alpha_1 = 10^6$ and $\alpha_2 = 1$ are taken as the best fitness.

10.5 Numerical Study of Controlled Adjacent Buildings with MR Dampers

This study investigates the efficacy of optimal semi-active dampers for achieving the best results in seismic response mitigation of adjacent buildings connected to each other by magnetorheological (MR) dampers under earthquakes, as shown in Fig. 10.2 and Fig. 10.5. The modified Bouc-Wen models with interactive relationships between damper forces and input voltages are used for MR dampers. One of the challenges in the application of this

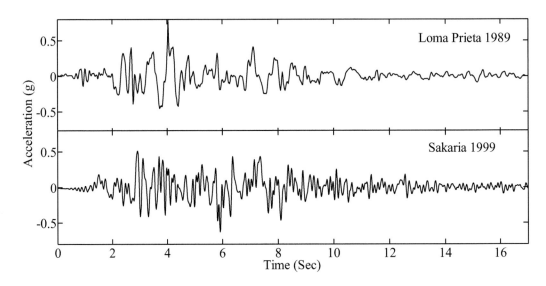

Fig. 10.5 Loma Prieta 1989 and Sakarya 1999 NS earthquakes with maximum ground accelerations scaled to 0.79 g and 0.51 g, respectively.

study is to develop an effective optimal control strategy that can fully utilise the capabilities of MR dampers. Hence, a significant task based on GAs is improved to obtain the optimal input voltages and number of dampers to understand the desired control forces at each floor level. LQR controller is used for obtaining the desired control forces, while the desired voltage is calculated based on Clipped Voltage Law (CVL). The control objective is to minimise the number of MR dampers installed between the buildings and both maximum displacement and drift storey responses of the buildings.

The multi objectives are first converted to a fitness function to be used in standard genetic operations, i.e. selection, crossover and mutation. The optimal control strategy generates an effective control system by powerful searching and self-learning–adaptive capabilities of GA.

10.5.1 Evaluation Criteria

The fitness of each individual can be obtained according to the defined function after determining the real value of each design variable in the population. This function reflects the desired objective. The control objective is to minimise both the peak displacement and peak drift responses of the structure to ensure the safety of the building and maintain the comfort level of the occupants. A set of evaluation criteria are based on those used for buildings to evaluate the various control algorithms given by Jansen and Dyke (1999, 2000). In this study, three of those criteria which are also regarded as the objectives in GA are selected to evaluate the efficacy of the proposed method. The first evaluation criterion is a measure of the normalised maximum floor displacement relative to the ground or the peak drift, given as

$$J_1 = \max_{t,i} \left(\frac{|x_i(t)|}{x^{max}} \right) \text{ or } \max_{t,i} \left(\frac{|d_i(t)|}{d^{max}} \right) \tag{10.11}$$

where $x_i(t)$ is the relative displacement of the *i*th floor over the entire response, and x^{max} denotes the uncontrolled maximum displacement response; $d_i(t)$ is the inter-storey drift of the *i*th floor $(x_i - x_{i-1})$, which is normalised by the peak uncontrolled floor drift denoted as d^{max}. The other evaluation criterion is the total number of MR dampers installed between the buildings as given by

$$J_2 = N_d \tag{10.12}$$

where $N_d = \sum_{j=1}^{m} n_j$, n_j is the total number of MR dampers at *j*th storey; m is the number of storeys of the lower building. The last evaluation criterion J_3 is the resulting γ in the H_∞ norm to be determined for the parameters of damper c_d, k_d because of installation of MR dampers. For simplicity, the objective function that reflects the above objectives in this study is defined as follows:

$$J = \alpha_c J_1 + \frac{(1-\alpha_c)}{2}(J_2 + J_3) \tag{10.13}$$

where J_1, J_2 and J_3 are the evaluation criteria representing normalised maximum floor displacement relative to the ground or normalised peak floor drifts, total number of dampers and damper parameters respectively; α_c is a weighting coefficient reflecting the relative importance of the three objectives. In this study, α_c is taken as 0.8. Each individual in the fitness function calculates the fitness of each individual. The positive fitness function is needed in GA, the problem of minimisation is converted such that the fitness has a positive value. Then the objective value is converted to fitness value given by

$$F = (C_p/J) \times \alpha \tag{10.14}$$

where C_p is a proper constant to make sure the fitness value is positive; α is a penalty constant to scale the fitness function. In this study, C_p and α are taken as 1 and 10, respectively (Yan and Zhou 2006, Arfiadi and Hadi 2011). This fitness function forms the basis of genetic operations in this study. The following procedures for fitness function have been investigated in this study.

10.5.2 GA-LQR and GA-H₂/LQG Optimisation Procedures

The GA function built in the MATLAB numeric computing environment is integrated into the SIMULINK block to simulate either LQR or H_2/LQG controller. Furthermore, binary coding is used to optimise the semi-active device system by defining the regulated output to obtain the required voltage and number of dampers for each floor level. Firstly, an LQR algorithm with full state feedback is employed in this study. For designing an LQR controller, the aim is to minimise the quadratic performance index:

$$J = \frac{1}{2}\int_0^\infty \left[x^T\, Q\, x + F_{mr}^T\, R\, F_{mr} \right] d_t \tag{10.15}$$

Using the LQR design, the optimal gain matrix K can be obtained for all X state vectors of the adjacent buildings. Thus, the desired force for MR dampers can be obtained. Secondly, a H_2/LQG controller is used as in Eq. 10.16.

$$J = \lim_{\tau \to \infty} \frac{1}{\tau} E\left[\int_0^\tau \left[y_m^T Q y_m + F_{mr}^T R F_{mr} \right] d_t \right] \tag{10.16}$$

Using the H_2/LQG design, the optimal gain matrix $K_c\,(s)$ can be obtained for all $\hat{X}$ measurements of adjacent buildings. In this study, some controllers evaluated by Dyke et al. (1996a) are used. The positive semi-definite weighting matrix Q and the positive definite weighting matrix of the performance index, which are given in Eq. 10.15 and Eq. 10.16 are taken as suggested by Chang and Zhou (2002).

$$Q = \mathrm{diag}\begin{bmatrix} K & M \end{bmatrix}$$
$$R = r\Lambda^T K\Lambda \tag{10.17a–b}$$

where r is the weight parameter reflecting the relative importance of reduction in the state vector X, which is taken as 0.75 in this numerical example.

10.5.3 Clipped Voltage Law Control in SIMULINK with Combined SOGA

The model of adjacent buildings based on structural control algorithms is implemented in SIMULINK, using continuous time systems. Numerical integration is conducted using the fourth-order Runge-Kutta solver of SIMULINK. After determining the ideal optimal force required for MR damper in the primary controller, the input voltage of the MR damper is determined in the secondary controller. This technique is referred to as clipped optimal control (Dyke et al. 1996a, Jansen and Dyke 2000). A graphic illustration of the clipped optimal control strategy is given in Fig. 10.6. The uncontrolled and controlled SIMULINK simulations are run simultaneously in this study. Using Eqs. 10.18a–b, the desired force can be calculated for LQR and H_2/LQG controllers. In other words, the primary controller is designed using these controllers.

$$f_d = -B^T\, R^{-1}\, PX = -K\,X$$
$$f_d = L^{-1}\left\{ -K_c\,(s)L\left(\begin{bmatrix} y_m \\ F_{mr} \end{bmatrix} \right) \right\} \tag{10.18a–b}$$

where f_d is the optimal desired force vector consisting of the element of f_{di}, that is the desired force for n_j number of MR dampers at the *j*th storey by the primary controller. As can be seen in Fig. 10.6, after checking dissipativeness based on the desired and actual damper forces in the secondary controller (see Eq. 4.18), continuously varying the input voltage for dampers at storey in the range of $[0 - V_{max}]$ can be stated in the secondary controller based on $H(.)$; heaviside step function expressed as 0 or 1.

A MATLAB based on SIMULINK model was built to simulate the system, including the MR damper model. The performance of MR damper is compared under seven controllers including two proposed methods (FLC combined GA and directly GA). They are namely passive off, passive on, semi-active controllers based on LQR and H_2/LQG, FLC combined GA (GAF) and directly GA. Under passive off and on strategies, MR dampers work as a passive device with damper command voltage set at zero and maximum (3 V, 6 V and 9 V), respectively. Under semi-active controllers, the damper command voltage is governed by control law based on LQR and H_2/LQG norms, whereas, under the proposed FLC combined GA strategy the fuzzy controller is used to determine the damper command voltage using the top-floor displacements of both buildings as inputs. Lastly, under another

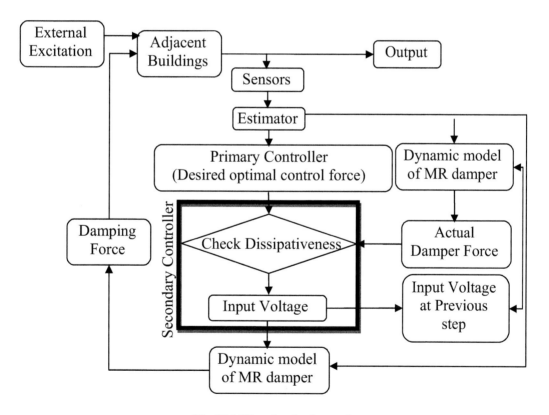

Fig. 10.6 Clipped optimal control.

proposed controller, the command voltage is directly determined based on the fitness function in GA. A comparative study is also conducted for maximum command voltage values of 2.25 V (only FLC controller case), 3 V, 6 V and 9 V.

Figure 10.7 shows a block diagram of the clipped optimal semi-active control system. Feedback for the controller is based on displacement measurements. A decision block based on manual switch can be added to the SIMULINK model as shown in Fig. 10.7, Fig. 10.8 and Fig. 10.9. This block switches the signal from the clipped-optimal control to the passive states. Figure 10.8 shows the simulation of Eqs. 10.19a–c for mechanical model of the MR damper in SIMULINK where the inputs are current and displacement and output is force. An MR damper model with a maximum capacity of 1000 kN is used in this study. The MR damper parameters have been suitably scaled to suit the damper deformation behaviour and the values of which are shown separately in the following numerical examples:

$$f_{mr}^i = c_1 \dot{y}_i + k_1 \left(x_{i+n} - x_i - x_0 \right)$$

$$\dot{y}_i = \frac{1}{\left(c_0 + c_1 \right)} \left\{ \alpha z_{di} + c_0 \left(\dot{x}_{n+i} - \dot{x}_i \right) + k_0 \left(x_{n+i} - x_i - y_i \right) \right\} \qquad (10.19a–c)$$

$$\dot{z}_{di} = -\gamma \left| \dot{x}_{n+i} - \dot{x}_i - \dot{y}_i \right| z_{di} \left| z_{di} \right|^{n_d - 1} - \beta \left(\dot{x}_{n+i} - \dot{x}_i - \dot{y}_i \right) \left| z_{di} \right|^{n_d} + A_c \left(\dot{x}_{n+i} - \dot{x}_i - \dot{y}_i \right)$$

The vectors of *ndii* and *vdi* in Fig. 10.7 represent the number of MR dampers and the current voltage for each storey, respectively. The strategies used in this study are carried out for two damper locations, namely, Case I: all floors of the lower building are connected with MR dampers to the adjoining floors of another building. In this case, five MR dampers per every floor are used. The same input voltage based on different control strategies is used for five MR dampers per every floor. In other word, all elements in the vector of *ndii* are equal to five. Case II: only alternative floors determined by GA of the lower building connected with the adjoining floors of another building. Further, the influence of damper command voltage is also examined for three sizes—3 V, 6 V and 9 V.

10.5.4 *Description of Model*

A system of buildings located adjacent to each other and interconnected by MR dampers possesses optimal semi-active control strategies as shown in Fig. 10.14. Building *A* is a 20-storey shear building discussed in Bharti et al. (2010) and Ok et al. (2008).

A 10-storey building discussed in Kalasar et al. (2009) and Pourzeynali et al. (2007) is taken as Building *B* which represents a typical medium size multistorey building in Table 10.10.

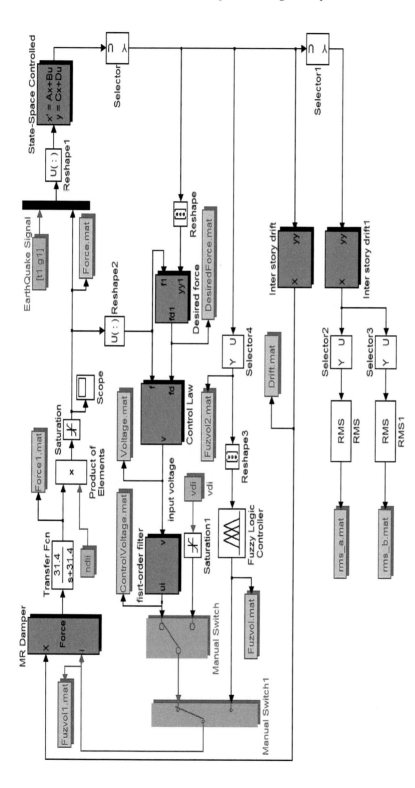

Fig. 10.7 Block diagram of semi-active control system, using H_2/LQG controller.

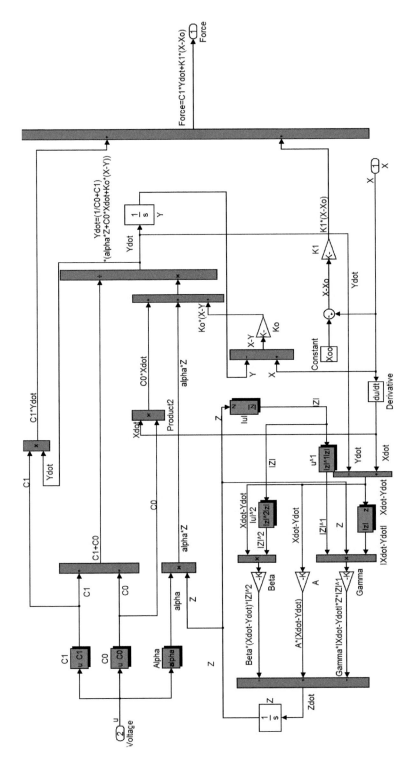

Fig. 10.8 Schematic diagram of the MR damper in SIMULINK.

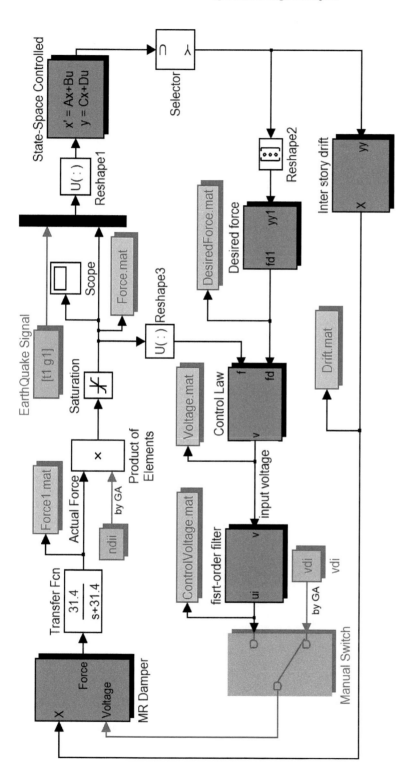

Fig. 10.9 Block diagram of semi-active control system using passive on controller.

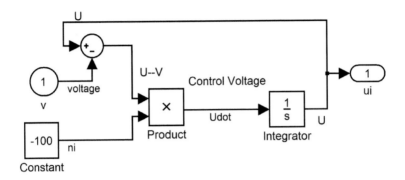

Fig. 10.10 Block diagram of first-order filter.

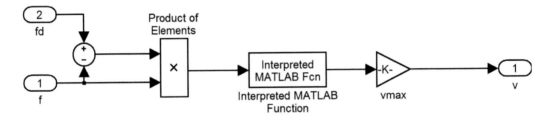

Fig. 10.11 Block diagram of control law.

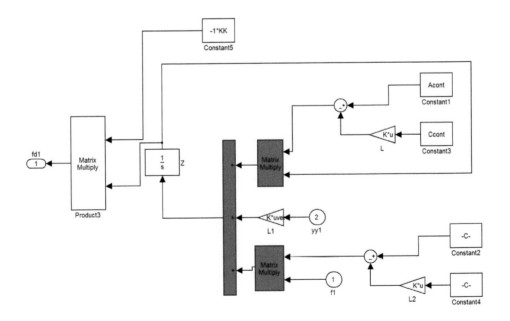

Fig. 10.12 Block diagram of desired force.

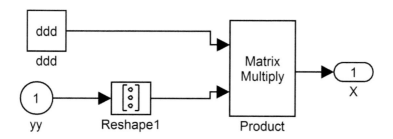

Fig. 10.13 Block diagram of desired force.

The structural parameters having mass, stiffness and damping coefficient are shown for both buildings in Table 10.10. Parameter variables that were obtained by Spencer et al. (1997) by optimal fitting of their model to test data are given in Table 10.11 and used in this book. The system is subjected to four earthquake ground motions—El-Centro 1940, Kobe 1995, Sakarya 1999 and Loma Prieta 1989, as shown in Fig. 10.2 and Fig. 10.5. In this study, the 20th and 10th floor displacements are conducted for two input variables of fuzzy logic control and the output variable is the command voltage sent to the MR damper. The MR damper parameters have been suitably scaled to suit the damper deformation behaviour and the values of which are shown in Table 10.11.

A stiffness proportional damping is assumed in Building A where the damping ratio of the fundamental mode equals about 5 per cent while Building B has also 5 per cent of damping ratio of the first mode.

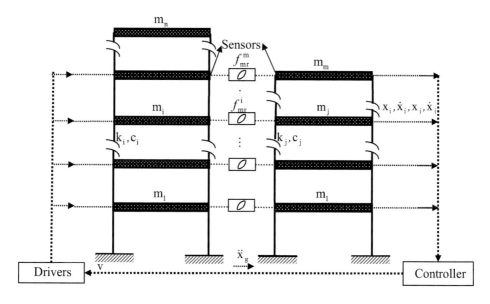

Fig. 10.14 N and M storey shear buildings with MR dampers.

Table 10.10 Structural Parameters of Both Buildings in Description of Model.

Floor (i)	Building A			Building B		
	m_i (t)	$K_i \times 10^6$ (kN/m)	$c_i \times 10^3$ (kN sec/m)	m_i (t)	$K_i \times 10^5$ (kN/m)	$c_i \times 10^3$ (kN sec/m)
1	800	1.4	4.375	215	4.68	1.676
2	800	1.4	4.375	201	4.76	1.648
3	800	1.4	4.375	201	4.68	1.585
4	800	1.4	4.375	200	4.5	1.585
5	800	1.4	4.375	201	4.5	1.539
6	800	1.4	4.375	201	4.5	1.539
7	800	1.4	4.375	201	4.5	1.539
8	800	1.4	4.375	203	4.37	1.539
9	800	1.4	4.375	203	4.37	1.099
10	800	1.4	4.375	176	3.12	1.146
11	800	1.4	4.375	–	–	–
12	800	1.4	4.375	–	–	–
13	800	1.4	4.375	–	–	–
14	800	1.4	4.375	–	–	–
15	800	1.4	4.375	–	–	–
16	800	1.4	4.375	–	–	–
17	800	1.4	4.375	–	–	–
18	800	1.4	4.375	–	–	–
19	800	1.4	4.375	–	–	–
20	800	1.4	4.375	–	–	–

Table 10.11 Parameters of Bouc-Wen Phenomenological Model Parameters for 1000 kN MR Dampers (Bharti et al. 2010).

Parameter	Value	Parameter	Value
c_{0a}	50.30 kN sec/m	α_a	8.70 kN/m
c_{0b}	48.70 kN sec/m/V	α_b	6.40 kN/m/V
k_0	0.0054 kN/m	γ	496.0 m^{-2}
c_{1a}	8106.2 kN sec/m	β	496.0 m^{-2}
c_{1b}	7807.9 kN sec/m/V	A_c	810.50
k_1	0.0087 kN/m	n_d	2
x_0	0.18 m	η	195 sec^{-1}

10.6 Summary

The numerical examples mentioned in this chapter were conducted for the two extreme cases, such as the pounding and the SSI systems. This study uses two groups of design variables of optimisation. They are the number of MR dampers installed at the first to the top floor of the lower building, the corresponding input voltage of the dampers. For these numerical examples, the earthquake records are shown. The design process uses the excitations of the NS components of the 1940 El Centro, the 1995 Kobe, the 1999 Sakarya and the 1989 Loma Prieta ground acceleration records. Numerical results of adjacent buildings controlled with MR dampers and the corresponding uncontrolled results are examined and compared with non-linear control algorithms. Chapter 11 shows the results obtained by the numerical examples as explained in the following pages.

11

Results and Discussion

11.1 Introduction

This chapter shows the results of both base-isolated coupled buildings and fixed-base adjacent buildings with applied equations of Chapter 2 and defined properties of numerical models in Chapter10.

11.2 Results of Base-isolated Coupled Buildings

This section consists of two main parts. The results of base-isolated buildings using direct integration method in MATLAB computer program with the capability of solving the equation of motion are shown in the longitudinal and transverse directions. The effect of pounding has using non-linear visco-elastic impact elements in this research is studied. For non-linear analysis, inelastic multi-degree of free lumped mass systems are modelled for structures and non-linear visco-elastic model for impact force during collisions are incorporated on the three-dimensional pounding between two adjacent four- and three-storey buildings.

11.2.1 Results of Response Analysis in the Longitudinal Direction

Figure 11.1, Fig. 11.2(a) and Fig. 11.3(a) show that after the first contact, Building *A,* which is lighter and more flexible than Building *B,* recoiled so significantly that it entered into the yield level at all storey levels (see Fig. 11.2(c), Fig. 11.3(c), and Fig. 11.4(b)). Though Building *A* does not have any contact during earthquake as shown in Fig. 11.1, it displayed the yield level in the first storey because of collisions at the second and third storey levels. Due to the fact that Building *B* kept small displacements, shear forces of Building *B* stayed in the elastic range. The results shown in Fig. 11.2(b) and Fig. 11.3(b) indicate that the most critical one for pounding problem is the highest contact point of buildings close to each other (at the third storey level) in view of the fact that contacts causing the maximum pounding force took place three times during the earthquake at this point. Results from the

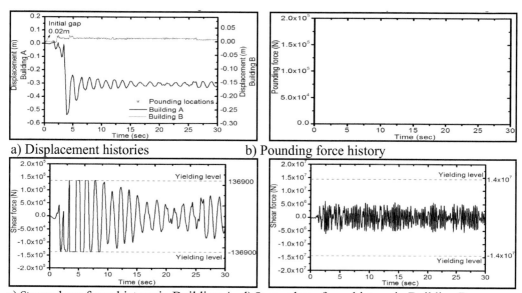

a) Displacement histories b) Pounding force history

c) Storey shear force history in Building A d) Storey shear force history in Building B

Fig. 11.1 Time histories in the longitudinal direction for the first storey of buildings.

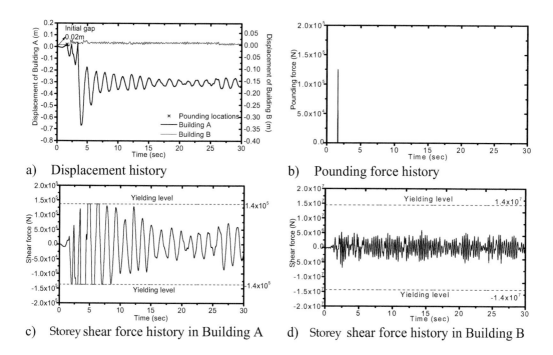

a) Displacement history b) Pounding force history

c) Storey shear force history in Building A d) Storey shear force history in Building B

Fig. 11.2 Time histories in the longitudinal direction for the second storey levels of buildings.

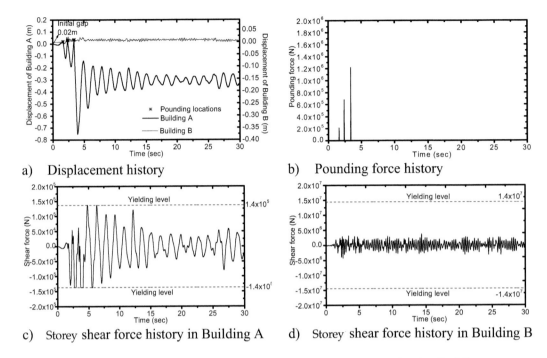

a)   Displacement history   b)   Pounding force history

c)   Storey shear force history in Building A   d)   Storey shear force history in Building B

Fig. 11.3 Time histories in the longitudinal direction for the third storey levels of buildings.

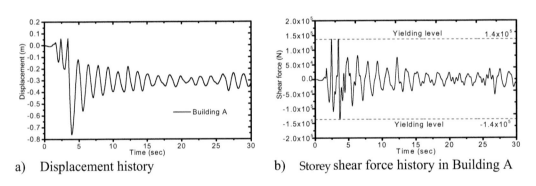

a)   Displacement history   b)   Storey shear force history in Building A

Fig. 11.4 Time histories for the top storey of Building *A* in the longitudinal direction.

current study for base-isolated adjacent buildings are compared with the results obtained from the study of Jankowski (2008) for fix-supported adjacent buildings. Figure 11.1 shows time histories in the longitudinal direction for the first storey levels of buildings.

Substantial permanent deformation of base-isolated buildings is seen as in Fig. 11.4.

However, the number of pounding and the values of pounding forces decrease in the ratio of one-third and 40 per cent, respectively when base isolation is used for the adjacent buildings.

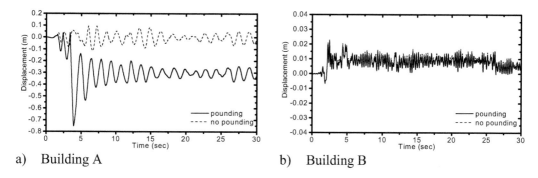

a) Building A   b) Building B

Fig. 11.5 Pounding-involved and independent vibration displacement time histories of the third storey levels of buildings in the longitudinal direction.

Entering into the yield range at all floors finally resulted in a substantial permanent deformation of the structure as can be seen in Fig. 11.5(a).

On the other hand, Building *B* (the heavier and the stiffer one) does not change at any considerable level in response to the earthquake-induced pounding between the structures (see Fig. 11.5(b)).

11.2.2 Results of Response Analysis in Transverse and Vertical Directions

The results of the study show that the responses of Building *A* in the transverse and vertical directions are considerably influenced by structure pounding, although this effect is not significant as in the case of longitudinal direction. It can be noted that collisions, in the results shown in Fig. 11.6(c) and Fig. 11.7(c), lead to yield level. Reaching yield level observed for all the floors in the transverse direction leads to structural deformation and even the amplitude of the deformation is not so large as in the longitudinal direction case. While shear forces for Building *A* are mainly the effect of ground motion excitation, the earthquake excitation is not enough to force the structure to enter the yield level for Building *B*.

On the other hand, the top storey of Building *A,* which has permanent deformation due to yield in the first three-storey levels, is influenced by the structural pounding as observed in Fig. 11.8(a).

11.3 Parametric Study of Base Isolated Buildings

A parametric study was conducted in order to determine the influence of different structural parameters on the pounding response of buildings. Building *A* is described as a reference building in Table 10.2. The results of parametric investigation carried out by changing the

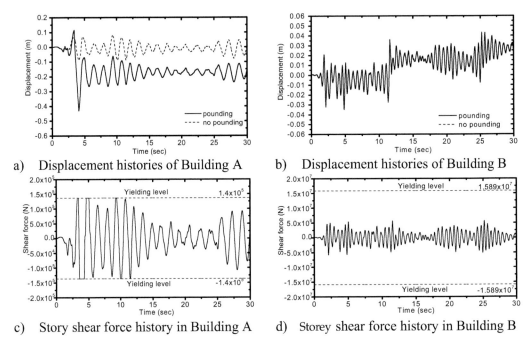

a)   Displacement histories of Building A   b)   Displacement histories of Building B

c)   Story shear force history in Building A   d)   Storey shear force history in Building B

Fig. 11.6 Time histories in the transverse direction for the second storey levels of buildings.

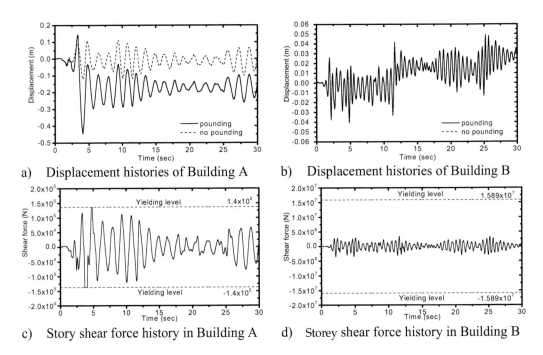

a)   Displacement histories of Building A   b)   Displacement histories of Building B

c)   Story shear force history in Building A   d)   Storey shear force history in Building B

Fig. 11.7 Time histories in the transverse direction for the third storey levels of buildings.

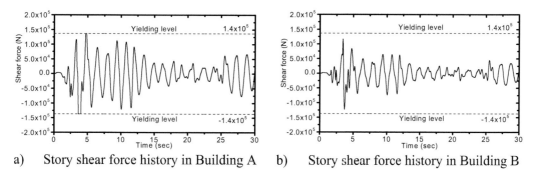

a)   Story shear force history in Building A   b)   Story shear force history in Building B

Fig. 11.8 Time histories in the transverse direction for the top storey of Building *A*.

values of structural parameters have been presented. For various values of gap between buildings, storey mass, structural stiffness, and friction coefficient of base isolation a numerical analysis has been carried out. When the effect of one parameter has been investigated, the values of others remained unchanged. For parametric analysis, the Duzce 1999 earthquake is used in Table 10.3.

11.3.1 Effect of Gap Distance

The peak displacements of the response in the vertical direction are similar to the transverse direction in almost all the ranges of the gap, mass, stiffness and the friction coefficient. Hence, they are not shown. In the case of longitudinal and transverse directions, an increase in the gap distance is associated with a reduction in the absolute displacement, although the peak displacement increases significantly in the lowest gap size values. As the gap size increases up to around 0.01 m, the absolute displacement also reaches the peak values. As can be observed in Fig. 11.9(b), there are no differences in the lowest gap size values. According to the results of a parametric study, a gap size of 0.12 m is required in order to prevent the pounding under Duzce 1999 ground motion.

Here, it should be highlighted that the minimum required distance between neighbouring buildings depends on both the dynamic characteristics of colliding buildings and the intensity of ground motion.

11.3.2 Effect of Storey Mass

The pounding response and the independent vibration displacement of the third storey of Building *A* in the longitudinal direction is shown in Fig. 11.10(c–d) with the storey mass $m_i = 1.4 \times 10^5$ kg corresponding to the peak pounding force in Fig. 11.10(a).

As can be seen in Fig. 11.10(a), the high value of pounding forces for the storey mass reaches up to about $m_i = 2.0 \times 10^5$ kg. Then, it drops and follows a steadily increasing slope.

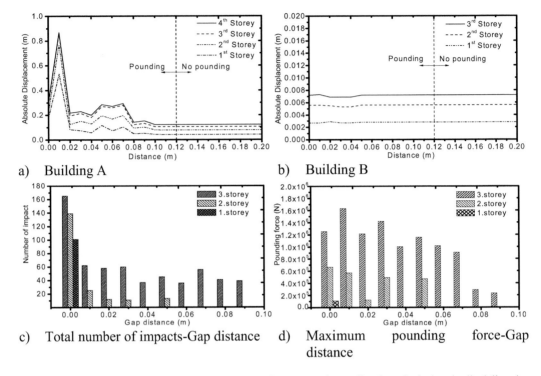

a)   Building A

b)   Building B

c)   Total number of impacts-Gap distance

d)   Maximum   pounding   force-Gap
distance

Fig. 11.9 Variation of peak displacement, the number of impacts and pounding force in the longitudinal direction in terms of the width of the gap between the buildings.

The pounding result in a significant change in the structural behaviour, including entering into the yield level, by comparing the pounding response and the independent vibration displacement of Building *A* in the longitudinal direction, as seen in Fig. 11.10(c).

11.3.3 *Effect of Structural Stiffness*

The independent vibration displacement and pounding response in the third storey of buildings are also illustrated in the longitudinal direction in Fig. 11.11(c–d) for the structural stiffness $k_i = 3.4 \times 10^6$ N/m corresponding to peak displacement as in Fig. 11.11(a). It can be seen from Fig. 11.11(a), in all the directions considered, the plots of the peak displacements differ greatly for Building *A*. In case of the longitudinal direction, the peaks have high values—in the vicinity of $k_{xi} = 3.4 \times 10^6$ N/m and $k_{xi} = 1.5 \times 10^7$ N/m. When comparing pounding response with independent vibration displacement of the third storey levels of the buildings, Fig. 11.11(c–d) indicates that pounding has a vital influence on the behaviour of both the buildings in the longitudinal direction.

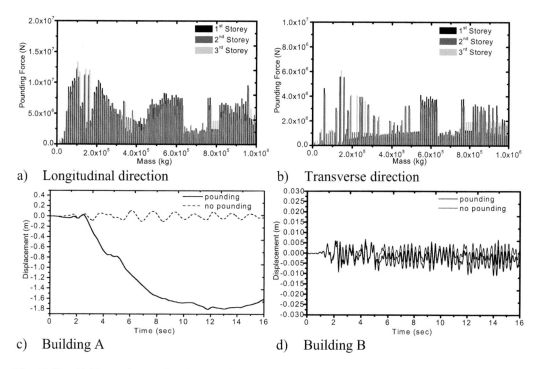

Fig. 11.10 a–b) The peak pounding force and storey mass in the longitudinal and transverse directions, c–d) pounding-involved and independent vibration displacement time histories of the third storey levels of buildings in the longitudinal direction for $m_i = 1.4 \times 10^5$ kg (i = 1, 2, 3, 4).

11.3.4 Effect of Friction Coefficient

The friction coefficient, $mu_a = 0.01$, corresponding to peak displacement in Fig. 11.12(a) in a plot of compression between the pounding-involved response and the independent vibration displacement is used in order to understand the effect of pounding on the behaviour of the buildings. It can be seen from Fig. 11.12(a, b) that the pounding-involved results of Building *A* have two ranges of a considered increase in the longitudinal and transverse directions till the parameter reaches to the vicinity of $mu_a = 0.13$.

The first one is around $mu_a = 0.01$, while the second is in the vicinity of $mu_a = 0.13$ in both directions. It can be seen from Fig. 11.12(c) that Building *A* enters into the yield level, even though Building *B* is nearly identical for the considered friction coefficient value as shown in Fig. 11.12(d).

11.4 Results of Fixed Buildings for SSI Effects

The pounding of adjacent buildings modelled as a non-linear response analysis of a three-dimensional MDOF numerical model as seen in Fig. 2.3 with either elastic or inelastic

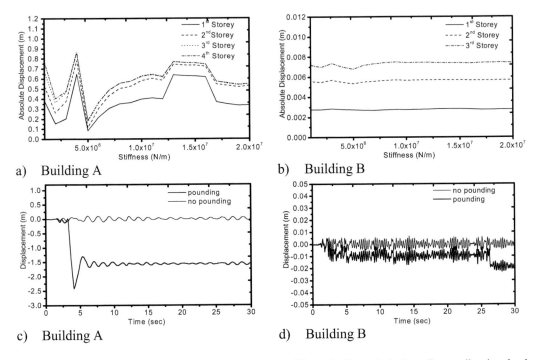

a)   Building A

b)   Building B

c)   Building A

d)   Building B

Fig. 11.11 a–b) Peak displacements with respect to storey stiffness, k_{xi} (i = 1, 2, 3, 4), c–d) pounding-involved and independent vibration displacement time histories of the third storey levels of buildings in the longitudinal direction for $k_{xi} = 3.4 \times 10^6$ N/m.

structural behaviour is studied under Case I or Case IV. Firstly, in order to investigate the SSI systems on the behaviour of coupled buildings with large and small SSI effects, the total response histories of Case I and Case IV are based on the deformation vectors of both superstructures of the two buildings modelled as elastic systems in Fig. 11.13 and Fig. 11.14, respectively. It can be seen in Fig. 11.13 that both buildings came into contact six times, based on the *x* directional displacements of the fourth floors during the earthquake.

Due to collisions in *x* direction and the effect of torque force, the contacts between the buildings in the *y* axis develop, although the pounding forces are not severe in comparison to the highest contact points in the *x* direction. While the lighter and more flexible Building *A* when compared with Building *B* is subjected to more twist about its *z* axes at the top floor levels at the lowest period of ground motion, the rotations in the top floor of Building *B* increase after the contact between the two buildings. It can be clearly noted from Fig. 11.14 that the number of contact points and the severity of pounding forces in both directions significantly increase between the buildings as modelled elastic systems. It shows the importance of SSI effects on the seismic response histories of adjacent buildings under the two-directional 1940 El-Centro earthquake. The roof twist of both buildings considerably decreases for Case I when compared to that of Case IV. Figure 11.15 and Fig. 11.16 show

216

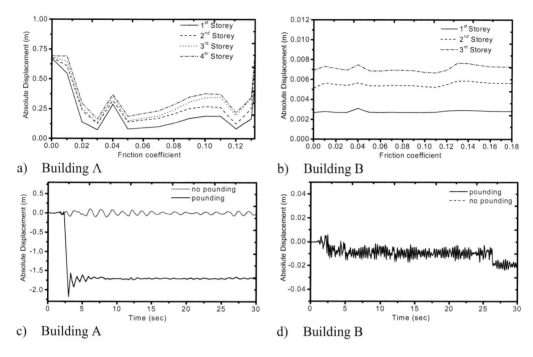

a)   Building Λ   b)   Building B

c)   Building A   d)   Building B

Fig. 11.12 a–b) Peak displacements with respect to friction coefficient, m_{ua}, c–d) pounding-involved and independent vibration displacement time histories of the third storey levels of buildings in the longitudinal direction for $m_{ua} = 0.01$.

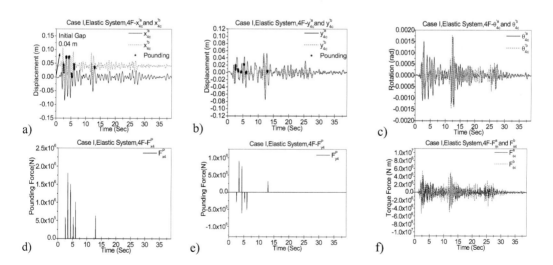

a)   b)   c)

d)   e)   f)

Fig. 11.13 Total response time histories of Case I on the fourth floors of adjacent buildings modelled as elastic systems under the 1940 El-Centro earthquake.

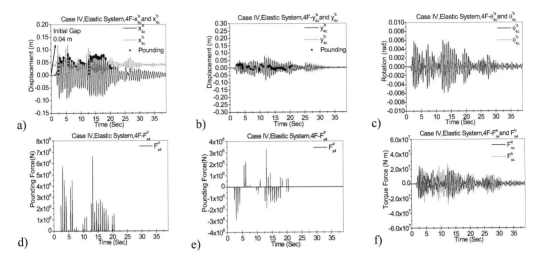

Fig. 11.14 Total response time histories of Case IV on the fourth floors of adjacent buildings modelled as elastic systems under the 1940 El-Centro earthquake.

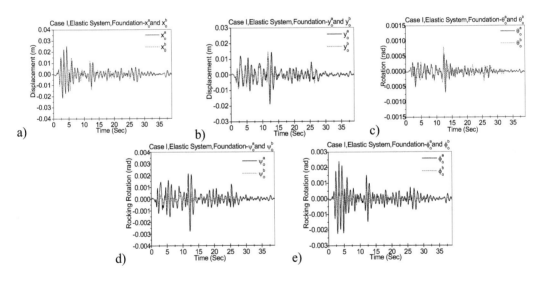

Fig. 11.15 Total response time histories of Case I at the foundations of adjacent buildings modelled as elastic systems under the 1940 El-Centro earthquake.

the deformation in translations, rotations in the x and y axes and twist about the vertical z axes and of the foundations of the buildings modelled as elastic systems, considering Case I and Case IV, respectively.

By comparing Fig. 11.15 and Fig. 11.16, it is seen that the values of the responses at the foundation for Case IV (hard soil) are considerably less due to the small SSI effect. In

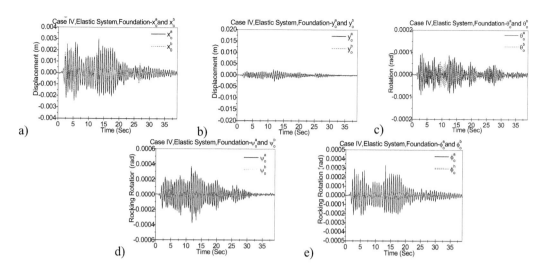

Fig. 11.16 Total response time histories of Case IV at the foundations of adjacent buildings modelled as elastic systems under the 1940 El-Centro earthquake.

order to investigate the effect of buildings modelled as either elastic or inelastic on the SSI effects, the deformation parameters with considerable pounding between the buildings and SSI forces are conducted here.

The total response histories of the superstructures of buildings that are modelled as inelastic systems are conducted under large and small SSI effects during the highest periods of the earthquake as seen in Fig. 11.17 and Fig. 11.18, respectively.

The response of inelastic systems is significantly different as compared to the response of the elastic systems. It can be seen that the values of peak pounding forces and the number of impacts are larger in elastic systems as compared with inelastic ones with large and small SSI effects. On the other hand, while the roof twist of both the buildings modelled as inelastic systems is almost similar to the elastic one for Case I, the response for Case IV in Fig. 11.18 become different compared with elastic systems. The heavier and stiffer Building *B* is subjected to the highest torque forces after collisions in the *x* and *y* axes under the earthquake by comparing as in Fig. 11.14. Moreover, in some cases, the lower storey levels of inelastic systems come into contact with each other, while in the case of elastic systems, collisions between the lower storey levels do not take place, as shown in Fig. 11.19.

Furthermore, the total response time histories, such as the storey shear forces in the *x* and *y* axes F_{xo} and F_{yo}, the storey torque force about the centre of resistance, $F_{\theta o}$, and the overturning moments in the *x* and *y* axes, $F_{\psi o}$ and $F_{\phi o}$ from rigorous analysis by the solution of Eqs. 2.22 and 2.26 is investigated to learn the effect of shear wave velocity with large and small SSI effects. Figure 11.20 shows the soil-structure interactions of Case I for coupled buildings

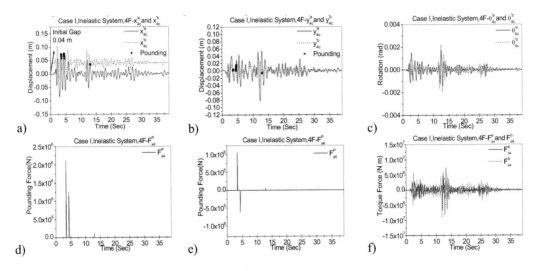

Fig. 11.17 Total response time histories of Case I at the fourth floors of adjacent buildings modelled as inelastic systems under the 1940 El-Centro earthquake.

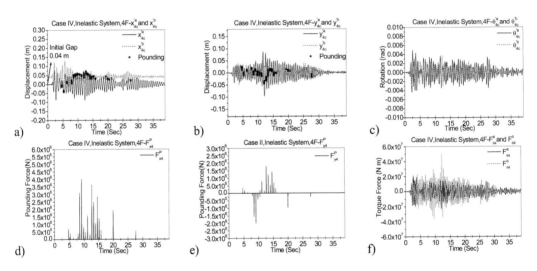

Fig. 11.18 Total response time histories of Case IV at the fourth floors of adjacent buildings modelled as inelastic systems under the 1940 El-Centro earthquake.

modelled as inelastic system for Case I. Figure 11.20 shows that when the shear wave velocity is high, the seismic response of coupled buildings modelled as inelastic system becomes considerably different at the foundation level due to the increasing soil-structure interaction forces. It can be found that the shear forces and torque force of Building *B*, that have the larger dimension of foundation as compared to Building *A*, suffered higher peak

220

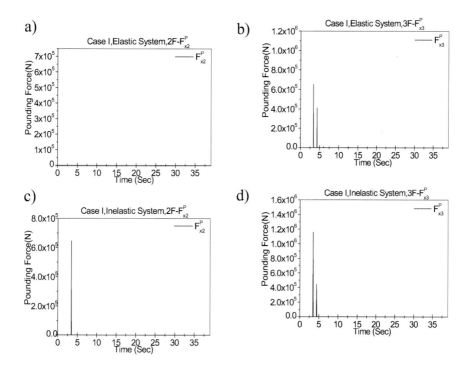

Fig. 11.19 Pounding force time histories under the 1940 El-Centro earthquake at second and third storey levels for the two buildings modelled as elastic and inelastic systems with a large SSI effect.

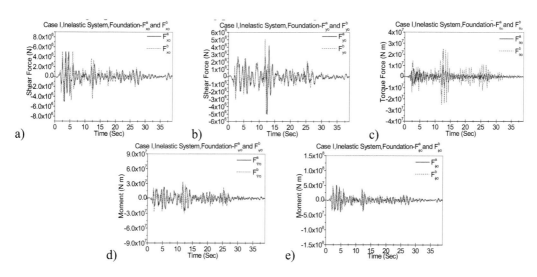

Fig. 11.20 Response time histories of the relative storey shears, torques and overturning moments of the foundation of coupled buildings modelled as inelastic system for Case I.

values after the first collision at the other storey levels. Figure 11.21 shows the response time histories of storey shears, torques and overturning moments based on the foundation of coupled buildings. Moreover, as can be estimated in Fig. 11.20 and Fig. 11.21, the values of overturning moment in the *x* axis become smaller than the overturning moment in the *y* axis due to larger plan dimensions being parallel to the longitudinal direction for each building.

On the other hand, while Building *B* is significantly affected, considering the small SSI effects, the lighter and more flexible Building *A* remains almost unaffected with peak values of torque force and overturning moment in the *y* direction.

The results obtained by the rigorous method show how the SSI affects the behaviour of adjacent buildings including the pounding effects. The approximate method, using MDOF modal equations of motion and the rigorous method using the direct integration method, are conducted. Note that the findings obtained by Balendra et al. (1982), Sivakumaran and Balendra (1994) and Jui-Liang et al. (2009) are compared with results obtained by the current study without consideration of the pounding effects, using MDOF modal equations of motion. In addition, the finding obtained by Jankowski (2006, 2008) is consistent with the results obtained by the current study using the direct integration method. Another four-storey asymmetric superstructure resting on a rigid base is modelled as the reference building (RB) in this study. The modal response of Building *B* for all cases is slightly less than but has the same trend with Building *A*. Hence, the results of Building *B* are not given here. Figure 11.22 shows the mode shapes of Building *A* resting on a rigid base and the mode shapes of all cases of Building *A* using the approximate method. Referring to Fig. 11.22, the offsets and slopes of thin lines represent the translations and rotations of

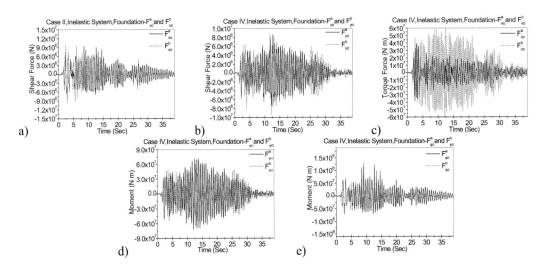

Fig. 11.21 Total response time histories of the relative storey shears, torques and overturning moments of the foundation of coupled buildings modelled as inelastic system for Case IV.

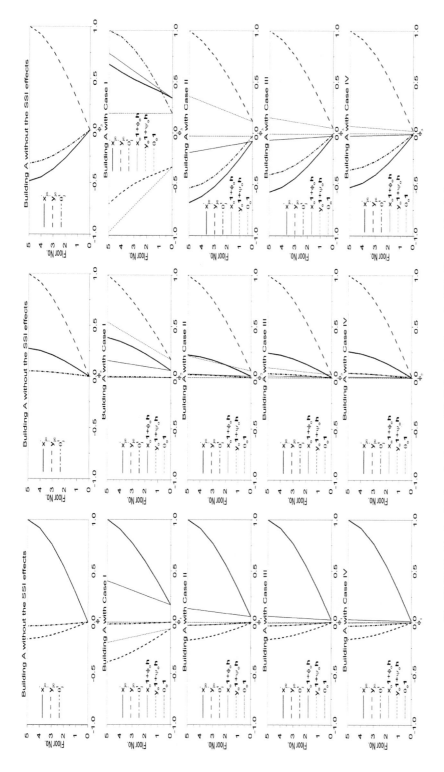

Fig. 11.22 First to the third mode shapes of Building *A* without the SSI effects and in all the cases.

the base. First to third mode shapes of dominant motion in both RB without the SSI effects and all cases are x-translation and y-translation directions in Fig. 11.22. However, in Case I, the dominant motion of the third mode shape is rocking in the y-direction. From Case I to Case IV, the first to third mode shapes become similar to the reference building. The modal displacement-time relationships of each eight degrees of freedom of Case I for each mode are clearly different as seen in Fig. 11.23. On the other hand, the modal responses of Case IV for the eight DOFs are similar to second and third modes. The modal responses of Case I for the third mode are mostly affected due to the SSI effects in conjunction with the phenomenon of out-of-phase vibrations. Figure 11.24 and Fig. 11.25 illustrate the total response histories of the extreme Case I and Case IV for both the approximate method (App.), using the MDOF modal equations of motion as in Eq. 2.36 and the rigorous method (Rig.) using the equation of motion for the whole SSI system in as Eq. 2.26 under the 1940 El-Centro and the 1995 Kobe earthquakes, respectively. While not only peaks but also the phase of total response histories of Case I and Case II are in agreement for both the methods in the selected earthquakes, the response histories of both buildings at the fourth storey in both methods are slightly different due to increased values of pounding force and the number of pounding under the small SSI effects as seen in Fig. 11.24 and Fig. 11.25.

It is observed from Fig. 11.24 and Fig. 11.25 that the number of poundings and the value of impact force are higher under the 1995 Kobe earthquake than the results of the 1940 El-Centro earthquake. Further, while increasing the SSI effects from soft soil to hard soil (Case I to Case IV), the number of poundings and the impact force increase under both the considered earthquakes.

Figure 11.26 shows modal response time histories without pounding of the first mode for Case IV. It can be seen from Fig. 11.26 that the modal responses of the eight DOFs for each vibration mode of Case IV are similar when the required distance is provided between the buildings to avoid pounding.

Figure 11.27 shows the displacement responses at the foundation of both the buildings under the two proposed methods used in this study during the 1940 El-Centro and the 1995 Kobe earthquake scaled at 0.8 g.

The responses at the foundation for Building *A* obtained by rigorous method in Fig. 11.28 and Fig. 11.29 are not in agreement with the results obtained by the approximate method in Case IV for both the considered earthquakes because of high pounding forces in the first to third modes. The responses at the foundation are less due to the small SSI effects when compared with the results in the large SSI effects.

However, it is interesting that in the large SSI effect (Case I), the responses in the translation, twist and rocking directions are same for both the methods under the 1995 Kobe earthquake. Figure 11.30 shows the displacement responses of Building *A* and Building *B* under Case I and Case IV and compared without considering the SSI effect.

The results in Fig. 11.30 show that displacement responses on the 4th floor of both the buildings under Case IV are higher than Case I. In order to understand how the pounding

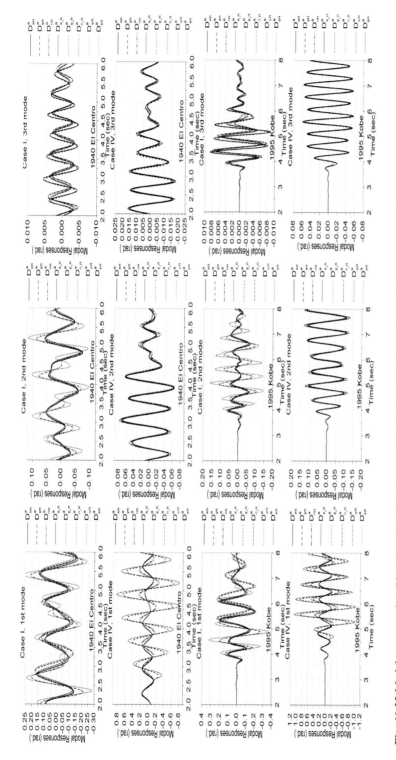

Fig. 11.23 Modal response time histories of the first to third modes for Case I and Case IV under excitation of the 1940 El Centro and the 1995 Kobe earthquakes.

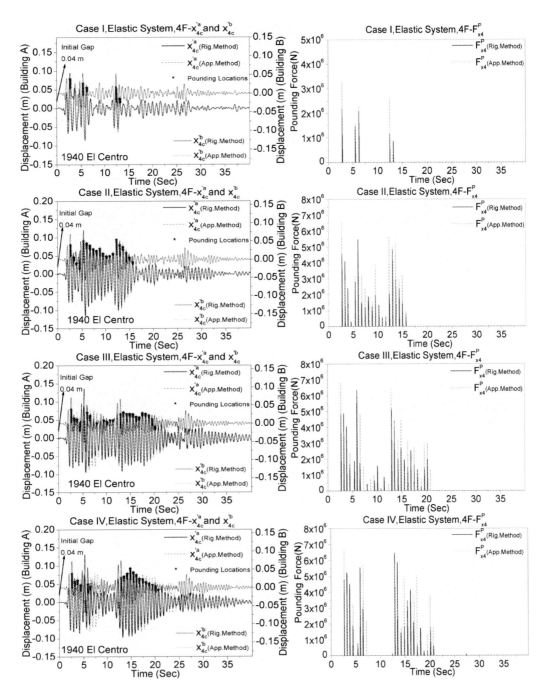

Fig. 11.24 Total response time histories of all the cases on the fourth floors under the excitation of the 1940 El-Centro earthquake for both approximate and rigorous solutions.

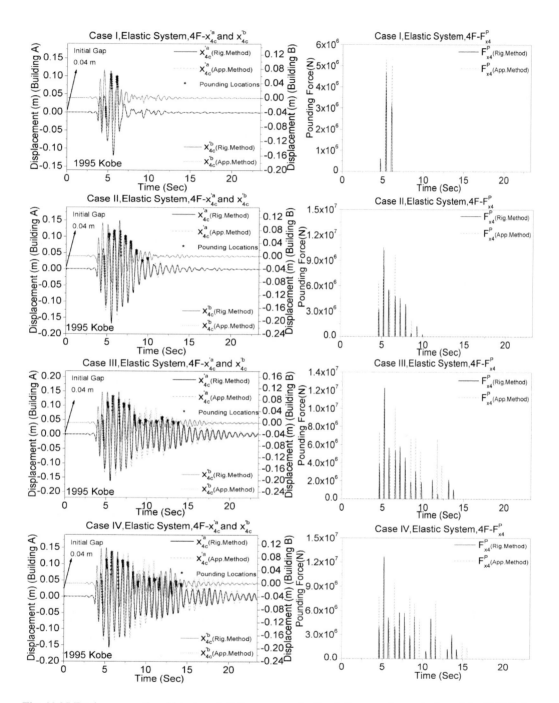

Fig. 11.25 Total response time histories of all the cases on the fourth floors under excitation of the 1995 Kobe earthquake for both approximate and rigorous solutions.

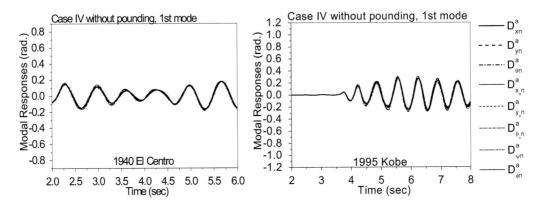

Fig. 11.26 Modal response time histories without pounding of the first mode for both Case IV under excitation of the 1940 El Centro and the 1995 Kobe earthquakes.

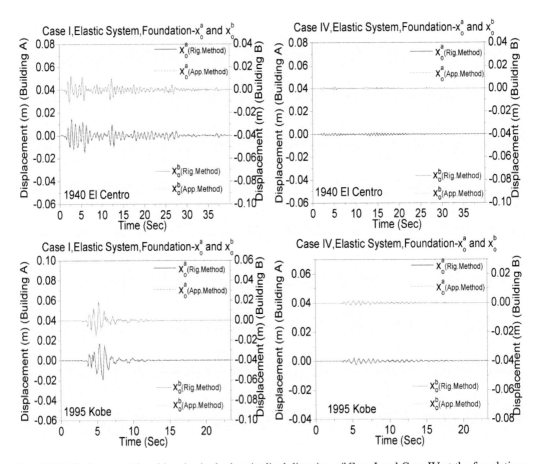

Fig. 11.27 Displacement-time histories in the longitudinal direction of Case I and Case IV at the foundations under the 1940 El-Centro and 1995 Kobe earthquakes.

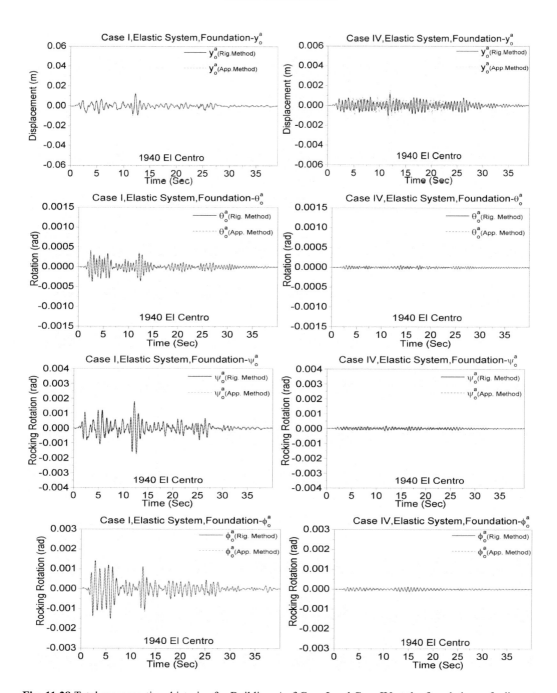

Fig. 11.28 Total response time histories for Building *A* of Case I and Case IV at the foundations of adjacent buildings under the 1940 El-Centro earthquake.

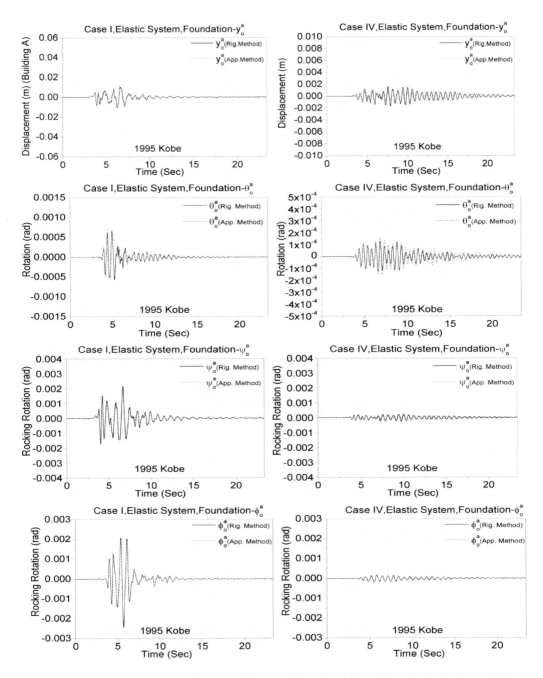

Fig. 11.29 Total response time histories for Building *A* of Case I and Case IV at the foundations of adjacent buildings under the 1995 Kobe earthquake.

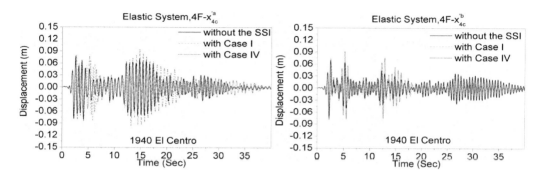

Fig. 11.30 Displacement responses of both the buildings without the SSI effects and compared to Case I and Case II under the 1995 El-Centro earthquake.

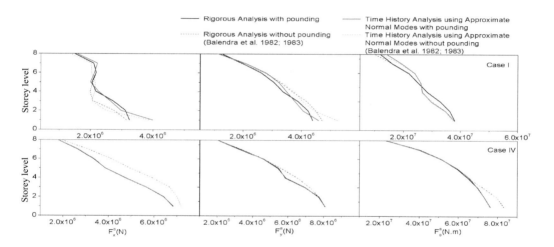

Fig. 11.31 Maximum storey, torques of 8 storey building-foundation system for Case I and Case IV with and without pounding effects under the 1940 El Centro earthquake.

affects the dynamic properties of the building under extreme SSI effects, Fig. 11.31 shows the maximum response of the system for the storey shears in the x- and y-directions and the storey torque about the vertical direction in shear wave velocities of 65 m/sec and 300 m/sec using numerical examples adopted by Balendra et al. (1982).

This is evident from the close agreement between the two solutions in Fig. 11.31 that the approximate and rigorous methods for adjacent buildings are accurate in the large SSI effects.

It is also seen from Fig. 11.31 that the maximum responses of the storey shear and torque due to the effect of pounding reduce when Case IV is used. The overall results show that the approximate method is slightly different than the results obtained by direct integration method in the small SSI effects due to the increase in pounding forces. The results in this study show that pounding affects the dynamic properties of the building under extreme SSI effects.

11.5 Results of Adjacent Buildings Connected with Control Systems

For enhancing the seismic performance of two adjacent buildings, the optimal analysis of both passive and active control systems was investigated in this research study. Two numerical examples were performed on i7-2630QM @2.9 GHz computer running MATLAB R2009a under two cases. The first example was adjacent buildings connected with passive damper systems; the second was adjacent buildings utilising active control systems. In this study, numerical examples' parameters used by Arfiadi (2000) were slightly modified. Furthermore, for passive control system, both binary and real coding were used and compared to optimise the passive device parameters. For active control system, real coded GA was used by defining the regulated output to obtain the controller gains. Several controllers were designed by choosing different combinations of measurements as feedback.

11.5.1 Results of Adjacent Buildings Connected with Viscous Dampers

Table 11.1 shows the damper parameters for all cases when different dampers are used in each floor level between buildings.

Table 11.1 Resulting Damper Parameters Obtained by Binary Coded GA-H$_\infty$ Optimisation.

Floors (i)	Parameters	Case A	Case B	Case C	Case D	Case E	Case F
1	c_{d1} (kN sec/m)	922.38	16.369	137.38	1995	13.126	127.28
	k_{d1} (kN/m)	2798.1	1606.7	3076.5	3620.5	3792.7	134.56
2	c_{d2} (kN sec/m)	93.17	656.45	0.383	43.23	7.48	98.53
	k_{d2} (kN/m)	2298.6	686.08	745.81	1990	2523.6	2934
3	c_{d3} (kN sec/m)	55.49	17.031	50.97	14.99	43.15	38.35
	k_{d3} (kN/m)	1742.4	578.11	1754.3	1617	1509.4	53.54
4	c_{d4} (kN sec/m)	17.69	89.031	5.85	20.59	135.4	187.26
	k_{d4} (kN/m)	785.32	251.77	188.98	369.01	435.69	296.06
5	c_{d5} (kN sec/m)	0.07	9.96	202.97	163.13	248.14	64.60
	k_{d5} (kN/m)	17.17	368.55	643.67	964.72	998.08	309
6	c_{d6} (kN sec/m)	349.13	234.36	217.61	156.05	109.51	216.57
	k_{d6} (kN/m)	1072.6	1998	1500	994.83	481.8	1891.8

As shown in Fig. 11.32, for every case, GA is run four times; Fig. 11.32 shows the evolving best fitness for Case A in H_2 optimisation.

For other cases the best fitness was same and is not shown here. In numerical Example 1, two design variables are chosen as stiffness k_d and the damping coefficient c_d when the same dampers are placed at all floor levels between buildings. For binary coded GA-H_2 and H_∞ optimisation, the lower and upper-bound values for each design variable are chosen. The lower and upper bound values of the stiffness are 0 and 4000 kN/m, while 0 and 2000 kN × *sec*/m are the lower- and upper-bound values for the damping coefficient, respectively. The length of the sub-chromosome (n bits) having three significant digit (p_i) for the stiffness is taken as 22, while the length of the sub-chromosome is taken as 20 for damping. It is noted that the optimisation results of the passive damper from H_∞ control are larger than the results from H_2 control. A reduction of about 25 per cent for Case A is between with and without controlled buildings as seen in Fig. 11.33.

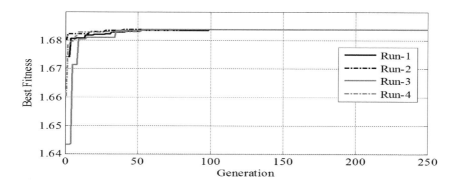

Fig. 11.32 Evolving best fitness for Case A with binary coded GA-H_2 optimisation.

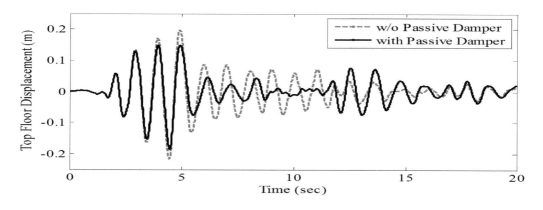

Fig. 11.33 Displacement response of the top floor of Building *A* for Case A with H_2 optimisation under El-Centro 1940 NS excitation.

In Fig. 11.34, comparison between optimised buildings utilising several regulated outputs is conducted. Table 11.2 shows the optimisation results of damper, damping and natural frequency of both buildings for all cases in H_2 control and Cases A–C in H_∞ control.

Reduction in the response is explicitly similar in all chosen regulated outputs either by using the optimised parameters from H_2 and H_∞ controls. The damper force is normalised with the weight of Building A in Fig. 11.35.

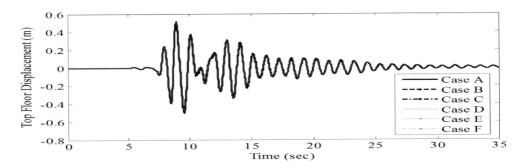

Fig. 11.34 Displacement response of top floor of Building B for all cases with H_2 optimisation under Kobe 1995 NS excitation.

Table 11.2 Optimisation Results of Numerical Example 1.

	Building A									
	H₂ (Binary Code)						**H∞ (Binary Code)**			**Un-controlled**
Component	**Case A**	**Case B**	**Case C**	**Case D**	**Case E**	**Case F**	**Case A**	**Case B**	**Case C**	
c_d (kN sec/m)	229.77	213.72	217.82	231.67	229.72	227.76	156.25	147.28	147.34	–
k_d (kN/m)	312.43	394.66	495.24	324.59	314.46	328.13	977.47	1078.20	1139.40	–
ω_1 (rad/sec)	4.98	5.03	5.09	4.99	4.98	4.99	5.33	5.37	5.40	6.29
ω_2 (rad/sec)	6.35	6.36	6.38	6.35	6.35	6.35	6.52	6.55	6.57	18.51
ω_3 (rad/sec)	13.24	13.26	13.28	13.24	13.24	13.24	13.40	13.42	13.43	29.65
ω_4 (rad/sec)	39.07	39.07	39.08	39.07	39.07	39.07	39.09	39.09	39.09	39.06
ω_5 (rad/sec)	46.22	46.22	46.22	46.22	46.22	46.22	46.23	46.23	46.24	46.21
ω_6 (rad/sec)	50.68	50.68	50.68	50.68	50.68	50.68	50.69	50.69	50.70	50.67
ξ_1 (%)	17.38	15.84	15.41	17.42	17.36	17.15	9.44	8.57	8.29	1.07
ξ_2 (%)	5.61	5.57	6.02	5.69	5.61	5.62	6.38	6.09	6.27	3.15
ξ_3 (%)	13.99	13.57	13.63	14.03	13.98	13.93	12.17	11.77	11.75	5.04
ξ_4 (%)	7.21	7.17	7.19	7.22	7.21	7.21	7.05	7.01	7.01	6.64
ξ_5 (%)	8.34	8.31	8.32	8.34	8.34	8.34	8.20	8.17	8.17	7.86
ξ_6 (%)	9.06	9.02	9.03	9.06	9.06	9.05	8.93	8.90	8.90	8.61

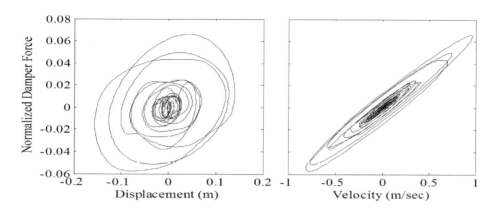

Fig. 11.35 Behaviour of fluid viscous damper under El-Centro 1940 NS excitation.

It can be seen from Fig. 11.35 that there is a significant energy dissipation in the passive case.

11.5.2 Results of Adjacent Buildings Connected with Active Viscous Dampers

A visco-elastic damper is placed on the top floor level between buildings with $c_d = 515.63$ kN × *sec*/m and $k_d = 3101$ kN/m. Note that the parametric values of visco-elastic damper are optimised by using the binary coding GA-H_∞ norm with Case F as the method in numerical Example 1. Figure 11.36 shows the evolving best fitness for Case F in H_∞ optimisation.

In numerical Example 2, the controller gains are obtained by using the real coded GA with the performance index H_∞ norm. One actuator and four gains are used in the numerical Example 2. Hence, four design variables are to be determined. In numerical Example 1, four runs are conducted by changing the initial value of the upper and lower bounds of the controller gain in each run. Hence, different values are conducted to explore the unknown domain of the gain and to see the robustness of the algorithm against the initial value. The first four runs of lower bound are chosen as -10, -10, 0 and -1000 respectively, while the first four run of upper bound are 10, -1, 100 and 100, respectively. Figure 11.37 shows evolving best fitness Example 2 Case F with inter-storey drifts and the top floor displacement of both buildings as the feedback and H_∞ norm as the objective function. Figure 11.38 and Fig. 11.39 show the response of the top floor of Building *A* Case F subject to El-Centro NS and Kobe NS excitations with maximum ground acceleration 0.3 g and 0.1 g, respectively.

It can be seen from Fig. 11.38 and Fig. 11.39 that the response is slightly reduced by using active systems, where the maximum control forces are $u_{max} = 2021$ kN and $u_{max} = 649$ kN for El-Centro earthquake with 0.3 g and 0.1 g, $u_{max} = 2107.5$ kN and $u_{max} = 721$ kN for Kobe earthquake with 0.3 g and 0.1 g, respectively.

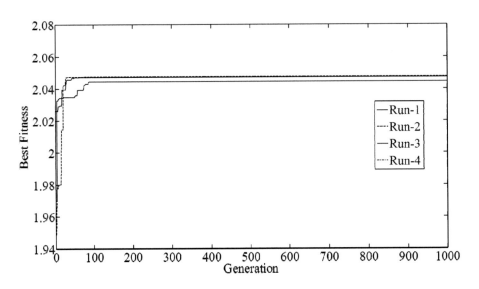

Fig. 11.36 Evolving best fitness for Case F with binary coded GA-H$_\infty$ optimisation.

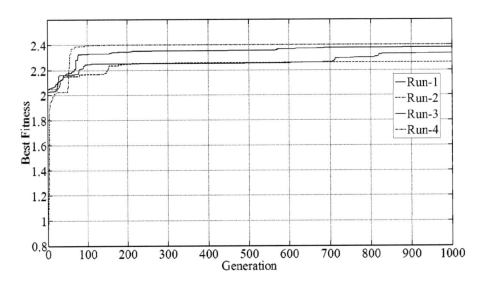

Fig. 11.37 Evolving the best fitness for the gains of Example 2 Case F with real coded GA-H$_\infty$ optimisation.

Figure 11.40 shows the transfer function from external disturbance to the top floor of Building *A*. The effect of linking buildings with either passive or active control systems compared with uncontrolled systems can be observed from a peak magnitude in *bode* diagram.

236

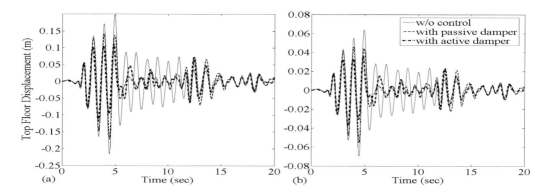

Fig. 11.38 Top floor displacement of Building *A* with uncontrolled, passive control and active control system due to El-Centro 1940 NS excitation (a) 0.3 g (b) 0.1 g.

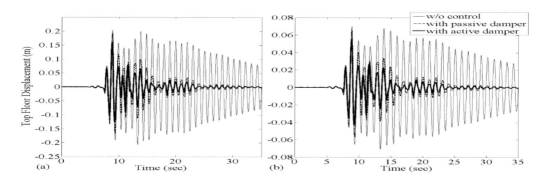

Fig. 11.39 Top floor displacement of Building *A* with uncontrolled, passive control and active control system due to Kobe 1995 NS excitation (a) 0.3 g (b) 0.1 g.

The designed adjacent buildings with dampers provide the reduction of response of each building around the frequency of the first mode, which is the highest contribution to response of each building as shown in Fig. 11.40.

11.5.3 Results of Adjacent Buildings Connected with MR Dampers

For numerical model in Section 10.5.4, the first nine natural frequencies of the 20-storey building for each example and the first five natural frequencies of the 10-storey building are given in Table 11.3. It is observed from Table 11.3 that the modes are well separated. The numerical example is subjected to four earthquake ground motions—El-Centro 1940, Kobe 1995 scaled to 0.8 g and 0.3 g, Sakarya 1999 and Loma Prieta 1989 in this study. The damper locations are investigated under two cases in this study, namely, Case I and Case II. In both cases, the responses are obtained with passive-off and passive-on cases which

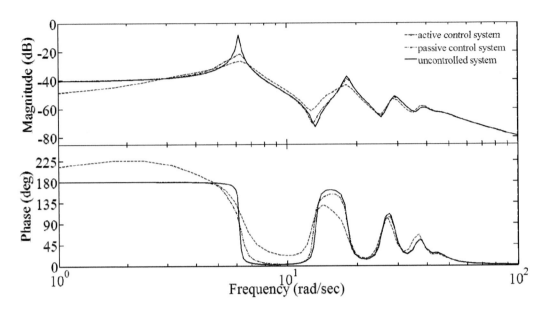

Fig. 11.40 Transfer function in Example 2 from external excitation to the top floor displacement of Building *A*.

Table 11.3 The Natural Frequencies of Model Examples Used for MR Dampers.

Frequency (Hz)	Building A	Building B
ω_1	0.510	1.148
ω_2	1.527	3.371
ω_3	2.535	5.443
ω_4	3.529	7.347
ω_5	4.501	9.075
ω_6	5.447	–
ω_7	6.361	–
ω_8	8.072	–
ω_9	8.860	–

are with constant zero voltage and with constant maximum applied voltage (i.e. 3, 6 and 9 V), respectively and compared with semi-active control cases based on LQR and H$_2$/LQG. In Case I, five MR dampers installed at each of the ten floors are considered for numerical example in Section 10.5.4. In Case II, only alternative floors of the lower building are connected with the adjoining floors of Building *A*. These alternative floors are determined by GA in Case II. In order to understand the influence of the command voltage required for MR damper, the study is examined for three magnitudes of V$_{max}$ (3 V, 6 V, 9 V).

Under passive-off (0 V), passive-on strategies (V$_{max}$ = 6 V) and semi-active based on LQR and H$_2$/LQG, the peak top floor displacement, the peak top floor acceleration, the peak storey shear and the peak base shear are examined as the response parameters of first interest. The base shear and storey shear of each building is regulated with the corresponding building weight and the damper forces are normalised with the weight of the lower building (Building *B*). Time variation of uncontrolled and the four controller responses of Building *A* and Building *B* corresponding to damper location of Case I is investigated. Figure 11.41 and Fig. 11.42 show the time response histories of top floor displacement of both buildings based on the considered four control strategies under the four different earthquakes and compared to the uncontrolled case. In time variation responses, the Kobe 1995 earthquake scaled to 0.3 g is used in order to compare explicitly with other earthquakes considered in this study.

In Fig. 11.41, passive-on and semi-active based on both the LQR and H$_2$/LQG norms result better than passive-off under El-Centro 1940 and Sakarya 1999 earthquakes while all control strategies have the same trend in Kobe 1995 and Loma Prieta 1989 ground motions. It is observed from Fig. 11.42, that all control strategies reduce the top floor displacement of Building *B* under all considered earthquakes. In terms of reduction of displacement responses, the performance of the control strategies in the lower building (Building *B*) is better than the higher building (Building *A*). Figure 11.43 and Fig. 11.44 indicate the time response histories of the top floor acceleration of Building *A* and Building *B*, respectively. The results in Fig. 11.43 indicate that for all control strategies, the overall trend is similar to the uncontrolled case in Building *A*. Figure 11.44 shows that semi-active controller based on H$_2$/LQG norm is effective in response mitigations for the lower building. The acceleration response reduction of Building *B* is higher under semi-active compared to passive-on strategy, except that under the 1999 Sakarya earthquake, semi-active has the same trends with passive-off and on strategies.

Although the response history of the top floor acceleration in the higher building is similar in passive-off and passive-on strategies, a comparative performance of the four strategies in Building *B* can be slightly observed in terms of acceleration responses. The response histories of the normalised base shear of both buildings are investigated in Fig. 11.45 and Fig. 11.46. The base shear of each building is normalised with the corresponding building weight. Therefore, the normalised base shear response of Building *A* is explicitly smaller than the normalised base shear response of Building *B*.

Further, Fig. 11.45 indicates that semi-active controllers are in agreement with the mitigation of base shear. All controllers show better performance compared to the uncontrolled case in Kobe 1995 and El-Centro 1940 earthquakes. Although the MR dampers work as passive devices with the maximum damper command voltage (6 V) under passive-on strategy, the response histories in terms of the normalised base shear in Fig. 11.45 almost match the uncontrolled case. It is observed from Fig. 11.46 that increase in base shear response is noted for Building *B* under passive-on strategy for Sakarya 1999 earthquake, while all control strategies exhibit better control performance for other three earthquakes. After the

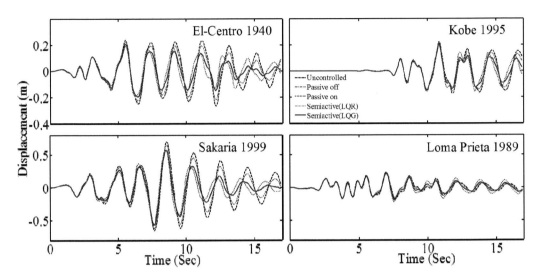

Fig. 11.41 Time response of top floor displacement of Building *A*.

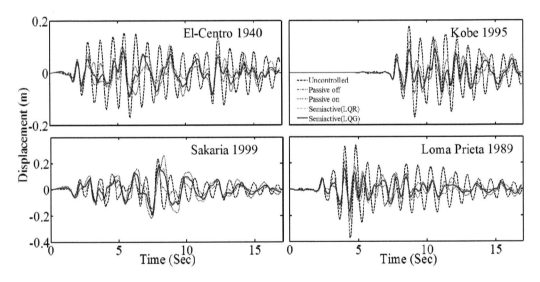

Fig. 11.42 Time response of top floor displacement of Building *B*.

comparative time history response plots, another comparative performance of the four control strategies is conducted in terms of peak floor displacement, acceleration and storey shear force based on the storey levels of both the buildings.

Figure 11.47 and Fig. 11.48 show the peak floor displacement of Building *A* and Building *B*, respectively.

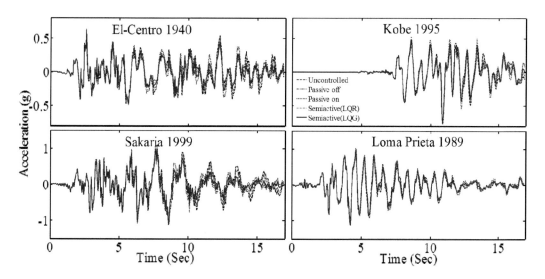

Fig. 11.43 Time response of top floor acceleration of Building *A*.

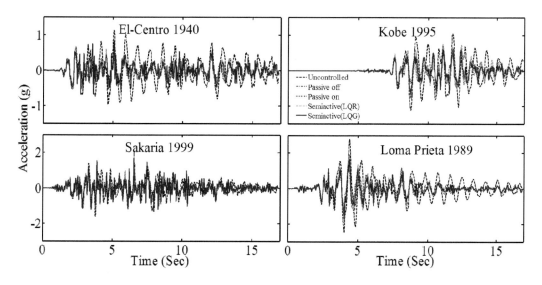

Fig. 11.44 Time response of top floor acceleration of Building *B*.

It is noted from Fig. 11.47 that the control performance of both passive-on and semi-active controllers is better than passive-off. Under passive-on strategy, peak floor displacements of Building *A* in Kobe 1995 earthquake are not good in terms of displacement reduction. The results of semi-active control strategies in Fig. 11.47 and Fig. 11.48 almost match for both buildings. Similarly, the overall trend in terms of semi-active controllers is similar

241

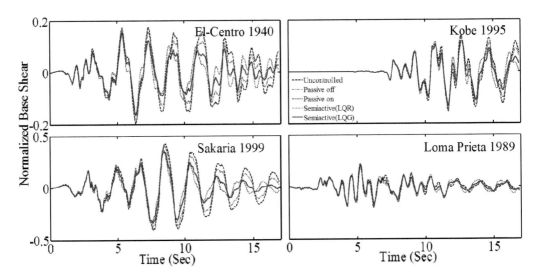

Fig. 11.45 Time response history of base shear of Building *A*.

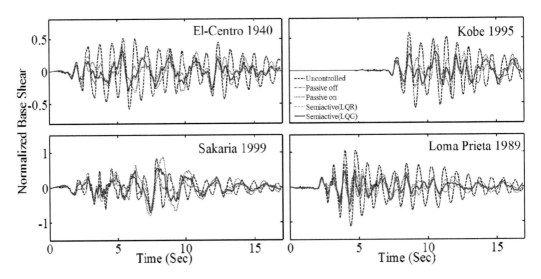

Fig. 11.46 Time response history of base shear of Building *B*.

for the lower building in Fig. 11.48. Passive-on strategy results in better displacement reduction under Kobe 1995 earthquake as compared to Building *A*.

Under Sakarya 1999 and El-Centro 1940 earthquakes, passive-off strategy provides the best reduction compared to semi-active controllers. The displacement response mitigation for higher floors of Building *B* is higher under semi-active compared to passive control

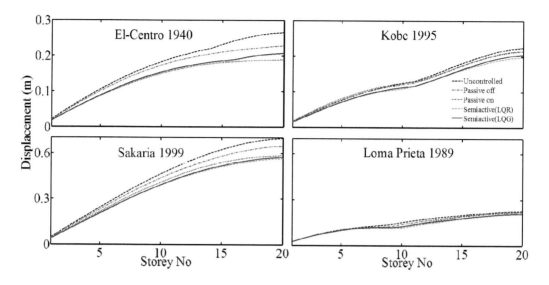

Fig. 11.47 Peak floor displacement of Building *A*.

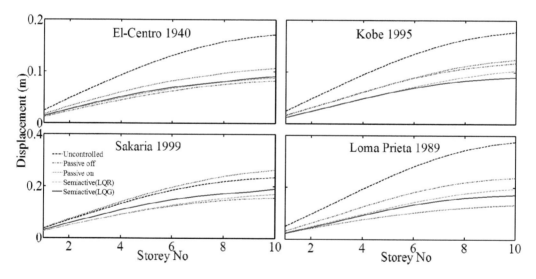

Fig. 11.48 Peak floor displacement of Building *B*.

strategies in the uncontrolled case. Figure 11.49 and Fig. 11.50 show the peak floor acceleration based on storey levels. It is observed from Fig. 11.49 that the control strategies do not show better results compared to the uncontrolled case based acceleration reduction for the higher building (Building *A*).

Increase in acceleration response is noted for higher floors under all control strategies for Kobe 1995 earthquake. On the other hand, all control strategies in acceleration response

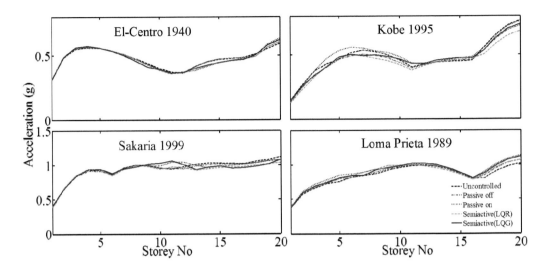

Fig. 11.49 Peak floor acceleration of Building *A*.

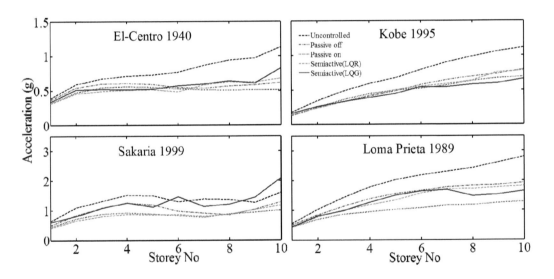

Fig. 11.50 Peak floor acceleration of Building *B*.

reduction for Building *B* are effective as depicted through Fig. 11.50. Semi-active controller based on LQR shows better mitigation that semi-active based on H_2/LQG for Building *B*. Acceleration reduction in the shorter building (Building *B*) is higher than the taller building (Building *A*).

Further, it is interesting that passive-on strategy in Fig. 11.50 shows a better response in terms of mitigation of the peak floor acceleration than semi-active and passive-off

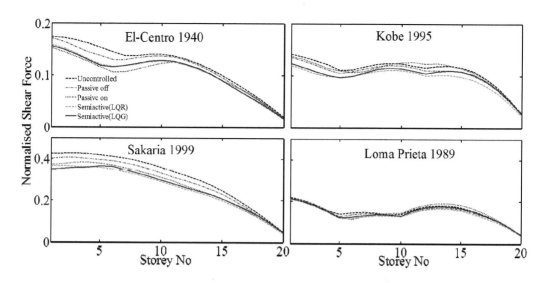

Fig. 11.51 Peak storey shear of Building *A*.

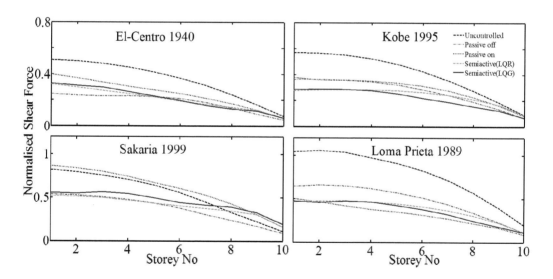

Fig. 11.52 Peak storey shear of Building *B*.

strategies. Figure 11.51 and Fig. 11.52 show the performance of the four control strategies in terms of storey shear for Building *A* and Building *B*, respectively. An overall best control performance is observed under semi-active controllers for all considered ground motion, especially in Sakarya 1999 for Building *A* and in Kobe 1995 for Building *B*. Passive-on strategy for Building *A* in Kobe 1995 and Building *B* in Sakarya 1999 is not effective to reduce the storey shear. For Building *A*, Kobe 1995 and Loma Prieta 1989 earthquakes show increase in storey shear with increasing storey levels. This is due to the fact that the sway of Building *A* is abruptly restricted by Building *B* as it suffers from high storey shear above tenth floor. Hence, this limitation results in increase of displacement response of Building *A* under passive-on strategy in Kobe 1995 and Loma Prieta 1989 earthquakes, as depicted in Fig. 11.47. In Fig. 11.52, reduction in response for Building *B* is observed under all considered earthquakes, except for Sakarya 1999 ground motion. Passive-on and semi-active controllers show better control of response as seen in Fig. 11.52. Hysteresis behaviour of MR damper under four control strategies, namely, passive-off, passive-on, semi-active-LQR and semi-active-H_2/LQG for the four different earthquakes is seen from Fig. 11.53 to Fig. 11.56.

It is observed that there is a significant energy dissipation in terms of displacement and velocity responses of MR damper in semi-active based on LQR and LQG norms as compared to passive-on ($V_{max} = 6$ V) and passive-off strategies. This study also investigated the influence of damper location and command voltage required for MR damper. In order to show the effectiveness of MR dampers, inter-connecting 10th floors of two buildings having different characteristics, the numerical model in Section 10.5.4 is used for the two damper locations and for three values of command voltage (3 V, 6 V and 9 V). The command voltages of MR dampers at each of the ten floors (Case I) between the buildings are determined by two methods proposed in this study.

Firstly, the optimal input voltage distribution of a fixed number of dampers is provided in this numerical example and compared to other control strategies. Before demonstrating the proposed multiple objective functions into a single-objective function, the performance of MR dampers is investigated by reducing the seismic response of adjacent buildings using various controllers. Firstly, five MR dampers are installed at each of the ten floors in the numerical example. All fifty dampers have the same input voltage.

Non-linear random vibration analyses by use of the 4th order Runge Kutta method is performed while varying the uniform input voltage from 0 to 10 V, which leads to variation in the damping capacity of the MR dampers.

Figure 11.57 and Fig. 11.58 show the maximum root-mean-square (r.m.s.) values of inter-storey drifts of the coupled systems by varying the uniform input voltage of MR damper under El-Centro 1940 and Kobe 1995 earthquakes, respectively. For decreasing the maximum inter-storey drift of the adjacent system, it is explained that an optimal value for the uniform input voltage of the MR dampers exists in a coupled structure system.

In this numerical example, the optimal input voltage of the MR dampers is 5.6 V for a uniform distribution of the 50-MR damper system in Kobe 1995 earthquake, while the optimal input voltage is 3.1 V in El-Centro 1940 earthquake. Use of the MR damper is important in damping capacity that can be easily adjusted by modulating the input voltage, without costly replacements or adjustments.

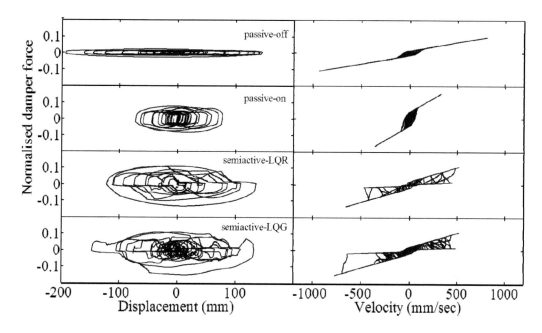

Fig. 11.53 The behaviour of MR damper under El-Centro 1940.

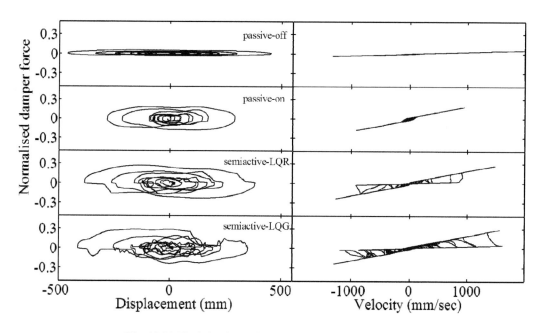

Fig. 11.54 The behaviour of MR damper under Sakarya 1999.

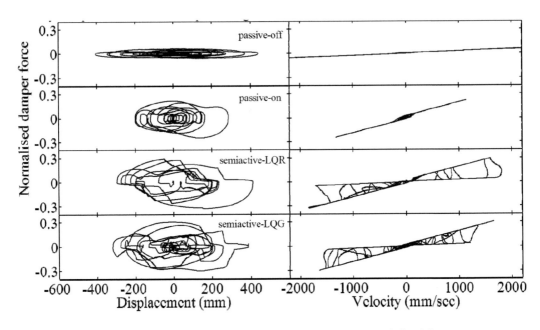

Fig. 11.55 The behaviour of MR damper under Kobe 1995 scaled to 0.8 g.

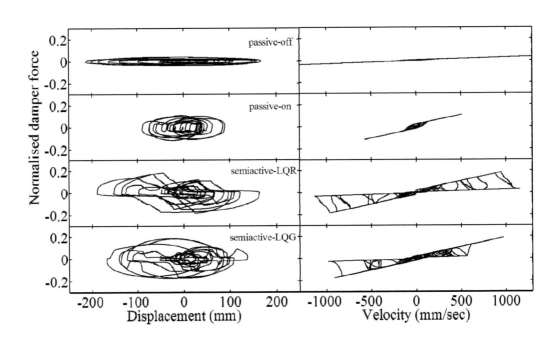

Fig. 11.56 The behaviour of MR damper under Loma Prieta 1989.

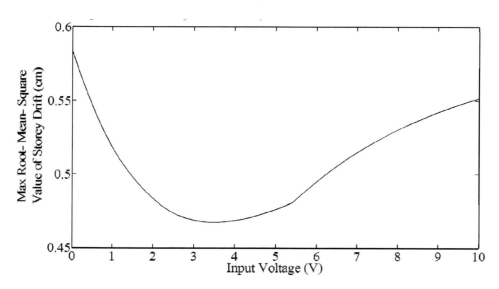

Fig. 11.57 Control performance of MR dampers with uniform input voltages under the 1940 El-Centro earthquake.

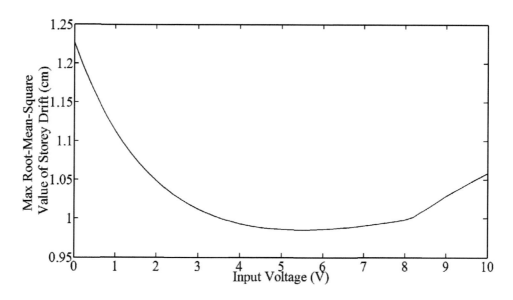

Fig. 11.58 Control performance of MR dampers with uniform input voltages under the 1995 Kobe earthquake scaled to 0.8 g.

In other words, varying the input voltages of the dampers is feasible in order to achieve an optimal performance. Hence, the results of the peak top floor displacement, acceleration and normalised base shear of the adjacent buildings using the optimum uniform voltage

(OUV) is evaluated with other control strategies used in this study. It is noted that varying the uniform input voltage *vdi* from 0 to 2.25 V for each storey is also conducted by means of the rule-based in fuzzy logic control in GA. Table 11.4 and Table 11.5 show the fuzzy rules generated by GA.

The results of GA optimisation are shown and a fuzzy logic controller is identified for both multi-input single-output (MISO).

This rule base obtained by GA establishes the correlation between the selected displacement inputs and the command voltage sent to MR damper when $\alpha_c = 0.6$; x_{20} and x_{30} represent the top floor displacements of Building *A* and Building *B*, respectively. When $\alpha_c = 0.6$ in Eq. 10.13, the multi-objective problem degenerates into a single-objective problem. The proposed strategy has the capability to mitigate the displacement response significantly while keeping the storey drift response at a low level. For GA, the population size is taken as 80 members and the upper limit on the number of generations as 10. Based on an elitist model, in order to perform evolutionary operations, proportional selection and one-point crossover are chosen in FLC combined with GA. The results of the damper locations in Case I are shown from Table 11.6 to Table 11.9.

It is observed from Table 11.6 that the overall displacement response reduction in Case I with passive-on strategy is as much with semi-active controllers except to passive-off for Building *B* in El-Centro 1940 earthquake. Further, the results show that there is no necessity to provide high command voltage for MR dampers and significant displacement

Table 11.4 Fuzy Rule-base Generated by GA in Case I under El-Centro 1940 Earthquake.

x_{20} \ x_{30}	NL	NS	ZO	PS	PL
NL	L	L	L	L	M
NS	ZO	M	L	L	S
ZO	S	ZO	S	M	M
PS	M	S	L	M	M
PL	M	M	L	S	L

Table 11.5 Fuzzy Rule-base Generated by GA in Case I under Kobe 1995 Ground Motion Scaled to 0.8 g.

x_{20} \ x_{30}	NL	NS	ZO	PS	PL
NL	ZO	L	S	S	S
NS	M	S	L	L	L
ZO	S	S	S	ZO	M
PS	ZO	M	ZO	S	ZO
PL	S	ZO	S	S	M

Table 11.6 Peak Top Floor Displacement under Different Control Strategies for Case I.

Earthquakes	Buildings	UNC	Case I												
			Off	Passive-on			LQR – CVL			H_2/LQG – CVL			OUV	GAF
				3 V	6 V	9 V	3 V	6 V	9 V	3 V	6 V	9 V		
El Centro, 1940	A	26.6	22.9	20.1	19.0	19.8	21.8	21.2	20.7	21.6	20.9	20.4	20.2	21.9
	B	17.1	8.1	8.6	10.4	11.5	8.4	8.6	8.7	8.6	9.1	9.6	8.8	7.9
Kobe, 1995	A	61.3	59.4	57.4	58.0	59.2	56.0	55.2	54.8	57.4	56.7	56.3	58.0	58.8
	B	48.0	35.0	25.6	30.6	32.9	29.6	27.5	26.0	30.5	27.7	25.5	30.7	29.1

Note: Displacement indicated is in × 10 mm. UNC: Uncontrolled

Table 11.7 Peak Top Floor Drift Inter-storey under Different Control Strategies for Case I.

Earthquakes	Buildings	UNC	Case I												
			Off	Passive-on			LQR – CVL			H_2/LQG – CVL			OUV	GAF
				3 V	6 V	9 V	3 V	6 V	9 V	3 V	6 V	9 V		
El Centro, 1940	A	0.25	0.24	0.23	0.23	0.23	0.23	0.22	0.22	0.22	0.22	0.22	0.23	0.23
	B	0.41	0.26	0.23	0.28	0.31	0.29	0.37	0.43	0.33	0.41	0.51	0.32	0.33
Kobe, 1995	A	0.88	0.83	0.81	0.84	0.87	0.75	0.74	0.75	0.81	0.82	0.83	0.84	80.1
	B	1.54	1.19	1.11	0.98	1.07	0.84	0.99	1.21	0.84	0.88	1.02	1.35	1.03

Note: The drift inter-storey indicated is in × 10 mm

response control is possible with less voltage in Building *B*. Using the optimum uniform voltage (OUV) and fuzzy controller combined by GA (GAF), a significant reduction for both buildings is observed under El-Centro 1940 earthquake although these proposed methods are not effectives in both the buildings under the Kobe 1995 earthquake. In Table 11.7, the top floor drift inter-storey responses show that the percentage reductions for Building *A* under passive-on strategy (Case I, 3 V) as compared to the uncontrolled case

Table 11.8 Peak Normalised Base Shear under Different Control Strategies for Case I.

Earthquakes	Buildings	UNC	Off	Passive-on			LQR – CVL			H_2/LQG – CVL			OUV	GAF
				3 V	6 V	9 V	3 V	6 V	9 V	3 V	6 V	9 V		
El Centro, 1940	A	0.20	0.19	0.17	0.16	0.16	0.18	0.17	0.17	0.17	0.16	0.16	0.17	0.18
	B	0.57	0.28	0.35	0.40	0.40	0.29	0.30	0.31	0.29	0.32	035	0.35	0.28
Kobe, 1995	A	0.40	0.40	0.41	0.42	0.43	0.39	0.39	0.38	0.38	0.37	0.37	0.42	0.41
	B	1.56	1.10	0.71	0.90	1.00	0.88	0.80	0.76	0.94	0.83	0.74	0.88	0.89

Table 11.9 Peak Damper Force of all MR Dampers at Top Floor Level of the Lower Building under Different Control Strategies for Case I.

Earthquakes	Off	Passive-on			LQR – CVL			H_2/LQG – CVL			OUV	GAF
		3 V	6 V	9 V	3 V	6 V	9 V	3 V	6 V	9 V		
El-Centro, 1940	54	127	166	203	181	287	379	192	315	426	127	80
Kobe, 1995	126	346	499	590	420	692	955	402	637	864	476	254

Note: The damper force indicated in kN

are: 8.0 under both the earthquakes. For Building *B*, the corresponding response reductions are 43.9 and 27.9 for El-Centro 1940 and Kobe 1995 earthquakes, respectively. However, a marginal increase in response is seen under semi-active controllers (9 V) for Building *B* under El-Centro 1940 earthquake. In Table 11.8, the percentage reductions in peak normalised base shear under passive-on strategy (6 V) for both buildings are: 20 and 29.8 and under semi-active based on H_2/LQG the reductions are 20 and 44 with El-Centro 1940. The proposed GAF control strategy and the uncontrolled case show that the correlation established by GA is effective. For Building *B*, a significant reduction in the normalised base shear responses is obtained by last column in Table 11.8. It is explicitly seen from Table 11.9, than an increase in damper voltage increases the force of the MR dampers, which is obvious. Using the proposed GAF controller, the damper force of all MR dampers at the top floor level of Building *B* is kept at a low level as compared to the other strategies in Table 11.9 under both the considered earthquakes. In Case II, only alternate floors of the lower building connected with adjoining floors of the higher building are considered for MR dampers.

The number of MR dampers *ndii* at each storey levels is minimised by using GA for economical benefits. At the same time, the input voltages are determined on the basis of the optimal number of MR dampers for each storey.

The MR dampers at each storey have the same input voltage. The optimal input voltages for each floor level by SOGA and NSGA II optimisations are determined for Case II in the given example in this study. The number of MR dampers and the corresponding command voltages are chosen as the design variables. Figure 11.59 and Fig. 11.60 show the variable distribution of command voltages with corresponding optimal uniform distribution under the considered earthquakes. The multi-objective optimisation is conducted by NSGA II to obtain the Pareto-optimal solutions as shown in Fig. 11.61.

In order to provide optimal damper systems in terms of cost-saving, another objective function is selected as the number of MR dampers to be minimised. Hence, the max r.m.s. storey drift of coupled buildings and the total number of the MR dampers are the two objective functions, which are the vertical and horizontal axes in Fig. 11.61, respectively. It is noteworthy that the total number of MR dampers is explicitly reduced by means of improved control performance, using GA and NSGA II optimisations. In other words,

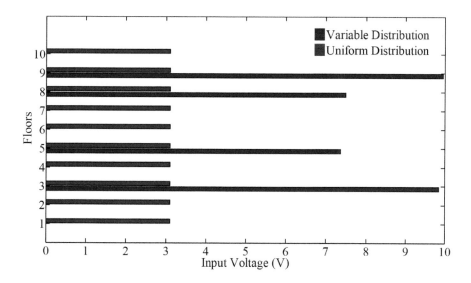

Fig. 11.59 Distribution of optimal input voltages of MR dampers under El-Centro 1940 earthquake.

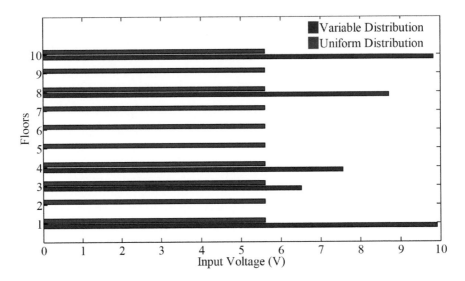

Fig. 11.60 Distribution of optimal input voltages of MR dampers under Kobe 1995 earthquake scaled to 0.8 g.

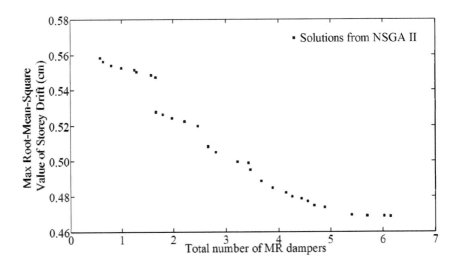

Fig. 11.61 The variation values of design variables under the 1940 El-Centro earthquake using NSGA II optimisation.

the performance efficiency of the damper system and cost effectiveness are considered in terms of the vertical and horizontal axes, though either use of more MR dampers or saving dampers causes damage to the buildings by means of the effect of cost or increase in the inter storey drift response of the two buildings. For this reason, this study achieves a similar level of seismic performance even with a significantly reduced number of dampers with the help of the proposed strategies. At the same time, input damper voltage is optimally determined. The maximum inter-storey drifts by uniform and varying voltages are 4.67 mm and 4.62 mm as seen in Fig. 11.57 and Fig. 11.61, respectively.

The optimum uniform voltage (OUV) is 3.1 V for El-Centro 1940 earthquake while for Kobe 1995, the OUV is 5.6 V under uniform distribution of the 50 MR damper systems. The total number of MR dampers are reduced from 50 to 6 for El-Centro 1940 and 8 for Kobe 1995 earthquakes as referred to in Fig. 11.62. When the number of MR dampers is reduced to 6 under El-Centro earthquake, the seismic performance of the coupled structural system is slightly mitigated and after that the response of max r.m.s. storey drift remains constant or increases slightly, as referred to in Fig. 11.61.

However, the total control voltage required to operate the MR dampers in the uniform distribution system turns out to be 155.0 V for El-Centro 1940 and 280.0 V for Kobe 1995 earthquake, whereas the variable distribution system uses a total of 52.1 V for El-Centro 1940 and 66.4 V for Kobe 1995 ground motion, as shown in Fig. 11.63.

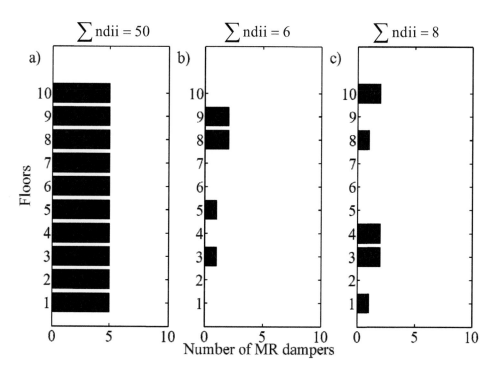

Fig. 11.62 Installation of MR dampers a) uniform distribution b) variation distribution in El-Centro 1940 and c) in Kobe 1995 earthquakes.

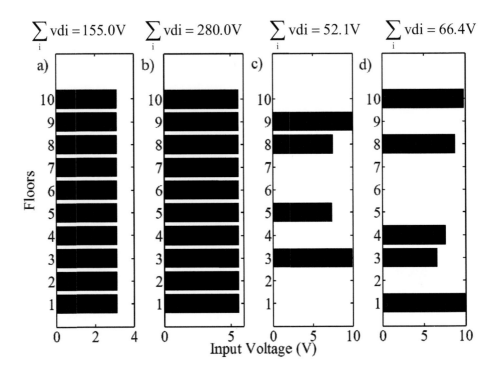

$\sum_i vdi = 155.0V$ $\sum_i vdi = 280.0V$ $\sum_i vdi = 52.1V$ $\sum_i vdi = 66.4V$

Fig. 11.63 Uniform distribution of a) El-Centro 1940 b) Kobe 1995, variation distribution of c) El-Centro 1940 and d) Kobe 1995 for input voltage of MR dampers.

The uniform distribution system, which has five MR dampers per every floor and uses the same input voltage, i.e. 3.1 V or 5.6 V for all the fifty dampers. The number of MR dampers and the variable distribution of the input voltage optimally are designed by a simple GA. The optimal system determined by the NSGA in this section is also denoted, which uses 6 or 8 MR dampers with optimal input voltage at each floor according to related earthquakes.

The response parameters of interest for adjacent buildings are peak top floor x_i, the peak floor drift d_i, the peak normalised base shear F_o and damper force F_{mr} as given in Table 11.10 and Table 11.13. Though the difference in the maximum inter-storey drift is not significant, the optimally designed variable distribution system uses less damping capacity with improved control performance in GA and NSGA II optimisations. Table 11.10 shows the peak top floor displacement of coupled systems under damper location Case II in terms of control strategies.

It is observed that numerical results given by the last column in Table 11.10, whose control strategy name is GA, shows the effectiveness of the newly generated number of MR dampers and the command voltage values. The percentage reduction of the drift inter-storey for Building *B* under passive-on strategy for Case II (6 V) in Table 11.11 as compared

Table 11.10 Peak Top Floor Displacement under Different Control Strategies for Case II.

Earthquakes	Buildings	UNC	Case II										
			Off	Passive-on			LQR – CVL			H_2/LQG – CVL			GA
				3 V	6 V	9 V	3 V	6 V	9 V	3 V	6 V	9 V	
El Centro, 1940	A	26.6	25.8	23.7	22.8	22.3	24.1	22.9	22.7	24.3	23.5	23.3	22.4
	B	17.1	13.8	9.4	8.06	7.75	10.9	10.3	10.2	11.9	11.0	10.4	7.79
Kobe, 1995	A	61.3	61.0	60.1	59.3	58.8	59.8	58.8	58.1	60.5	60.2	60.0	58.9
	B	48.0	45.9	39.8	35.2	31.4	42.1	40.1	38.7	42.7	40.1	38.0	31.5

Note: The displacement indicated is in × 10 mm

Table 11.11 Peak Top Floor Drift Inter-storey under Different Control Strategies for Case II.

Earthquakes	Buildings	UNC	Case II										
			Off	Passive-on			LQR – CVL			H_2/LQG – CVL			GA
				3 V	6 V	9 V	3 V	6 V	9 V	3 V	6 V	9 V	
El Centro, 1940	A	0.25	0.25	0.25	0.24	0.23	0.24	0.24	0.24	0.25	0.24	0.23	0.23
	B	0.41	0.46	0.37	0.35	0.34	0.40	0.40	0.41	0.40	0.39	0.43	0.34
Kobe, 1995	A	0.88	0.87	0.85	0.83	0.82	0.84	0.82	0.81	0.87	0.87	0.86	0.83
	B	1.54	1.49	1.40	1.34	1.30	1.41	1.43	1.48	1.44	1.41	1.39	1.32

Note: The drift inter-storey indicated is in × 10 mm

to the uncontrolled are 14.6 and 5.7 under El-Centro 1940 and Kobe 1995 earthquakes, respectively. The corresponding reductions under semi-active controller based LQR are 2.44 and 7.1. Similar trends are also observed under damper location Case II and command voltage 3 V and 9 V. This shows that the percentage reduction under passive-on is better than semi-active strategy. For damper location Case II, all control strategies reduce the base shear response of both the buildings in Table 11.12.

Table 11.12 Peak Normalised Base Shear under Different Control Strategies for Case II.

Earthquakes	Buildings	UNC	Case II										
			Off	Passive-on			LQR – CVL			H_2/LQG – CVL			GA
				3 V	6 V	9 V	3 V	6 V	9 V	3 V	6 V	9 V	
El Centro, 1940	A	0.20	0.19	0.19	0.19	0.18	0.18	0.18	0.18	0.18	0.19	0.19	0.18
	B	0.57	0.47	0.32	0.28	0.26	0.37	0.34	0.33	0.41	0.38	0.36	0.27
Kobe, 1995	A	0.40	0.39	0.40	0.40	0.40	0.39	0.39	0.39	0.40	0.39	0.39	0.40
	B	1.56	1.47	1.26	1.10	0.97	1.33	1.24	1.19	1.35	1.25	1.16	0.98

Table 11.13 Peak Damper Force of all MR Dampers at Top Floor Level of Lower Building under Different Control Strategies for Case II.

Response		Off	Case II										
			Passive-on			LQR – CVL			H_2/LQG – CVL			GA
			3 V	6 V	9 V	3 V	6 V	9 V	3 V	6 V	9 V	
F_{mr}	E	82	240	360	464	249	397	537	270	430	478	489
	K	176	578	858	1085	610	978	1341	624	997	1329	1177

Note: F_{mr} is the damper force indicated in kN; E: represents the 1940 El-Centro; K: represents the 1995 Kobe earthquake scaled to 0.8 g.

The overall base shear reduction for damper location Case I with optimal uniform input voltage is as much with damper location Case II with optimum command voltage obtained by GA and NSGA II optimisations. In Table 11.13, it is noteworthy that the peak damper force of all MR dampers increases as compared to Case I although the total number of MR dampers decreases. However, increase in the damper forces of Case II is observed because of increase in command voltage under all control strategies. This shows that connecting only at the alternative floors reduces the cost of dampers significantly by 80 per cent and response reduction is also explicitly observed. Increase in voltage causes increase in damper stiffness; therefore, attracts more force according to formulations as shown in Eqs. 10.19a–c and Fig. 10.8. Another numerical model is conducted in this study. The 20th and 10th storey buildings in Table 11.14 are used as same as structural parameters of Building *A* of the numerical model in Section 10.5.4.

Table 11.14 Structural Parameters of Both Buildings Having Same Characteristics.

Floor (i)	Building A			Building B		
	m_i (t)	$k_i \times 10^6$ (kN/m)	$c_i \times 10^3$ (kN sec/m)	m_i (t)	$k_i \times 10^5$ (kN/m)	$c_i \times 10^3$ (kN sec/m)
1	800	1.4	4.375	800	1.4	4.375
2	800	1.4	4.375	800	1.4	4.375
3	800	1.4	4.375	800	1.4	4.375
4	800	1.4	4.375	800	1.4	4.375
5	800	1.4	4.375	800	1.4	4.375
6	800	1.4	4.375	800	1.4	4.375
7	800	1.4	4.375	800	1.4	4.375
8	800	1.4	4.375	800	1.4	4.375
9	800	1.4	4.375	800	1.4	4.375
10	800	1.4	4.375	800	1.4	4.375
11	800	1.4	4.375	–	–	–
12	800	1.4	4.375	–	–	–
13	800	1.4	4.375	–	–	–
14	800	1.4	4.375	–	–	–
15	800	1.4	4.375	–	–	–
16	800	1.4	4.375	–	–	–
17	800	1.4	4.375	–	–	–
18	800	1.4	4.375	–	–	–
19	800	1.4	4.375	–	–	–
20	800	1.4	4.375	–	–	–

It may be noted that more MR dampers are required to provide better seismic control performance for coupled buildings with same structural parameters as compared to the numerical model in Section 10.5.4. Figure 11.64 shows multi-objective optimisation for both buildings with same structural parameters and performed by NSGA to obtain the pareto-optimal solutions.

It is observed that more MR dampers are required to reduce the structural responses of both buildings. Objective 1 represents the total number of MR dampers while Objective 2 shows max value of r.m.s. storey drift (cm) in Fig. 11.64. It can be seen from Fig. 11.64 and Fig. 11.65, that optimisation values of max value r.m.s. storey drifts with varying the command voltage for MR dampers at each storey levels with more than 50 number of dampers for both buildings having the same characteristics.

For these reasons, the damper location Case II for same buildings is not conducted in this study. Nevertheless, a comparative performance of five control strategies, namely, uncontrolled, passive-off, passive-on, semi-active controllers based on LQR and H_2/LQG,

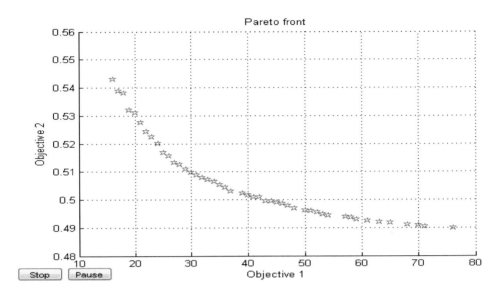

Fig. 11.64 Objective functions obtained by pareto front.

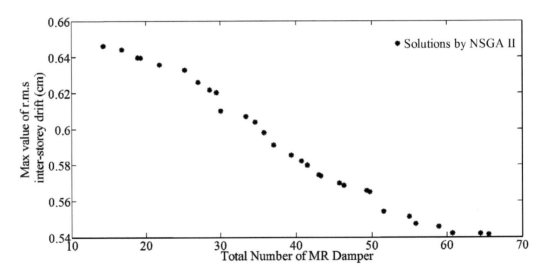

Fig. 11.65 Damper designs in the space of objective functions.

in terms of peak floor displacement, acceleration and storey shear force is conducted through Table 11.15 and Table 11.16 under both the earthquakes. In Table 11.15, the best percentage reduction in peak displacement of both buildings between controller strategies used is passive-on strategy with command voltage 6 V for El-Centro 1940 and 9 V for Kobe 1995 earthquakes.

Table 11.15 Peak Floor Displacement, Peak Floor Drift Storey, Peak Normalised Base Shear and Damper Force under Different Control Strategies during El Centro 1940 Earthquake.

Responses	Buildings	UNC	Case I									
			Off	Passive-on			LQR – CVL			H_2/LQG – CVL		
				3 V	6 V	9 V	3 V	6 V	9 V	3 V	6 V	9 V
x_i (× 10 mm)	A	26.6	22.8	20.2	19.0	19.3	21.0	20.8	20.3	21.4	20.7	20.2
	B	17.1	13.5	8.6	8.7	8.4	9.7	9.2	9.0	10.5	9.5	9.10
d_i	A	0.25	0.23	0.21	0.21	0.23	0.22	0.20	0.21	0.20	0.20	0.20
	B	0.44	0.37	0.29	0.27	0.26	0.31	0.30	0.31	0.37	0.38	0.39
F_o	A	0.20	0.18	0.16	0.15	0.14	0.17	0.17	0.16	0.16	0.16	0.15
	B	0.50	0.39	0.22	0.23	0.25	0.26	0.23	0.22	0.32	0.29	0.26
F_{mr} ($\sum F_{mr}$)	–	–	55 (273)	165 (825)	258 (1292)	316 (1578)	175 (877)	304 (1519)	422 (2109)	175 (874)	300 (1502)	435 (2173)

Table 11.16 Peak Floor Displacement, Peak Floor Drift Storey, Peak Normalised Base Shear and Damper Force under Different Control Strategies during Kobe 1995 earthquake.

Responses	Buildings	UNC	Case I									
			Off	Passive-on			LQR – CVL			H_2/LQG – CVL		
				3 V	6 V	9 V	3 V	6 V	9 V	3 V	6 V	9 V
x_i (× 10 mm)	A	61	60	55.1	52.9	51.8	55.9	55.2	54.5	57.2	56.2	55.5
	B	52	48	37.7	31.5	27.3	40.7	38.2	37.1	45.5	43.7	42.2
d_i	A	0.9	0.8	0.76	0.72	0.79	0.75	0.72	0.71	0.83	0.83	0.83
	B	1.4	1.3	1.04	0.99	0.99	1.11	1.04	1.06	1.29	1.28	1.27
F_o	A	0.4	0.4	0.38	0.36	0.35	0.37	0.37	0.36	0.36	0.35	0.35
	B	1.2	1.1	0.89	0.72	0.61	0.95	0.87	0.84	1.06	1.00	0.96
F_{mr} ($\sum F_{mr}$)	–	–	161 (805)	499 (2344)	665 (3326)	843 (4217)	498 (2492)	809 (4044)	1113 (5000)	584 (2920)	944 (4721)	1303 (5000)

11.6 Summary

Based on the results of base-isolated buildings in this research study, the pounding between symmetric buildings affects significantly the lighter and more flexible buildings associated with dynamic characteristics.

Based on the results of the fixed-base asymmetric buildings, the response of Building *A* at the superstructure is significantly affected due to the large SSI effects. Moreover, when the shear wave velocity is high, the seismic response of coupled buildings modelled on inelastic system become significantly different at the foundation level due to the increasing soil-structure interaction forces.

The multi-objective genetic algorithm deals with mutually conflicting objectives in terms of the number of dampers and the response of structures. As a result, increasing the number of dampers does not necessarily increase the efficiency of the system. In fact, increasing the number of dampers can prove worse for the dynamic response of the system. The proposed GAF and GA methods for the output voltage of MR dampers achieve enhanced seismic performance with economical efficiency. GAF and directly used optimum number of damper and command voltage obtained by GA show better control in displacement and damper force response of Building A although all the strategies are better for the shorter building (Building B).

12
Summary and Conclusions

This book presents some representative results that have been performed to evaluate the effects of base isolation on the seismic response behaviour of isolated adjacent buildings with significantly different dynamic properties in the time domain. Furthermore, the effect of soil-structure interaction (SSI) in multi-storey asymmetric adjacent buildings is investigated for impact cases, using the approximate method for MDOF modal equations of motion and the rigorous method for the direct integration method. The differential equations of motion for two-way asymmetric shear buildings are derived and solved by the fourth-order Runge-Kutta method with and without impact. The main objective of this study is to fill the gap in knowledge of the seismic response of adjacent buildings retrofitted by control systems, such as fluid viscous damper, active control devices and MR dampers in semi-active systems, using different adjacent building models. A comprehensive computational investigation is done in order to design optimum control devices between adjacent buildings. This book aims not only to reduce the seismic responses, but also to minimise the total cost of the damper system. From the numerical simulations conducted in this research, the following conclusions can be drawn:

- The results of the response analysis of base-isolated buildings demonstrate that pounding of the structures during ground motion excitation has a significant influence on the behaviour of lighter buildings in the longitudinal direction. This pounding may lead to substantial amplification of dynamic response, which may cause considerable permanent deformations in the neighbouring buildings

- The results of the parametric study explicitly show that the behaviour of the heavier building in the longitudinal, transverse and vertical directions is practically unchanged by pounding of structures

- In some cases, the lower storey levels of inelastic systems come into contact with each other, although collisions between lower storey levels in the case of elastic systems do not take place

- Modelling the colliding buildings to behave inelastically is really important in order to mitigate the effect of pounding on the behaviour of coupled buildings. The results of further investigation show that the responses based on deformation vectors of superstructures for each building are significantly reduced by increasing the shear wave velocity, while the SSI forces increase at the foundation of the buildings

- It is observed that the roof twist of the lighter building, which is assumed to be inelastic, decreases for the large SSI effect compared to the small SSI effect. At high shear wave velocity, the top floor deformations of couple buildings are slightly on the conservative side

- The results in this study show that pounding affects the dynamic properties of the building under extreme SSI effects. The values of pounding force in the approximate method remain same as the results obtained by the rigorous method when the shear wave velocity is very low, although those results are not in agreement with the small SSI effects due to increased number of poundings. The responses at foundation are less in the small SSI effects compared with the results in the large SSI effects. Instead of using only the first few vibration modes, all modes are used to achieve satisfactory results by the rigorous method due to high pounding forces

- The best-fitness individual is copied to the next generation by using an elitist strategy. The result shows that the visco-elastic damper is quite effective in reduction of about 25 per cent in the response of both the adjacent buildings. The response mitigation is better for the shorter building

- Static output feedback controller, where the gain is multiplied directly with the measurement output without an observer, is used as feedback control. It makes the system framework simpler than a dynamic output controller. As a result, adjacent buildings coupled by either passive or active damper at the top floor level demonstrate that the controlled design approach can systematically achieve enhanced seismic performance with economical efficiency instead of using damper at all the storey levels

- The control scheme of coupling two adjacent buildings using MR damper linkages is very effective for seismic response reduction in both the buildings. For coupled system the response control is better for the shorter building. A comparison of seven considered controlled strategies, namely the uncontrolled case, passive-off, passive-on, H_2/LQG, LQR, GAF and directly GA shows that the controlled strategy based on H_2/LQG is more effective than the other strategies. Significant response control is possible with passive-on strategy that provides the effectiveness of MR damper in response mitigation even if the control algorithm fails.

- A fuzzy controller combined with genetic algorithm is developed to regulate the damping properties of the MR damper and reduce the chosen different regulated outputs in multi-degree of freedom seismically-excited adjacent buildings. From the results obtained, it can be concluded that the fuzzy controller proposed with GA is successful in reducing maximum and drift storey responses of the selected buildings under three different earthquake excitations.

Based on the case study given detailly in Chapters 7–9, the time variation of the top floor displacement and base shear responses of two buildings linked by fluid viscous dampers at all the floors, with optimum damping coefficient and optimum damping stiffness, was shown. It can be clearly seen that dampers had a major role on the effectiveness of dampers in controlling the earthquake responses of both the adjacent buildings. The effectiveness of fluid viscous dampers became less important for the higher building than the lower and more beneficial for the adjacent buildings, having different heights than the same. Example 4(a) investigates to mitigate the seismic response of coupled buildings with different shear stiffness but same heights and finds that the results are more effective than the adjacent buildings having the same characteristics. For all the studied buildings in this study, the reductions in top floor displacement become important in N-S direction.

In order to reduce the cost of fluid viscous dampers, the response of adjacent buildings was investigated by considering only specific dampers (i.e. almost 50 per cent of the total) with optimum parameters shown above at selected floors. For lower buildings, lesser dampers at appropriate placements were more effective than dampers at all floors. However, for higher buildings, it can be the opposite in the case big earthquakes. Changing the parameters of dampers in terms of damping coefficients is important to reduce the top floor displacements of adjacent buildings. Damper 2, which has much damping coefficient, is more beneficial than Damper 1.

The fluid viscous dampers are found to be very effective in reducing the earthquake responses of adjacent linked buildings. There is an optimum placement of dampers for minimum earthquake response of two adjacent connected buildings. This study shows that it is not necessary to connect the two adjacent buildings with dampers at all floors but lesser dampers at select locations can reduce the response during seismic events. Although lesser dampers can be effective in reduction of the responses, the amount of reduction is higher than in other cases when the dampers are located at all floors. The study can be further explored by considering when two high adjacent buildings with different damping characteristics, are connected a new and different damper.

Furthermore, using the fluid damper of parameters derived based on two damping coefficients is beneficial to reduce the responses of the adjacent structures under select earthquakes. Further, the diagonal location of fluid viscous dampers causes reduction in displacement, acceleration and shear force responses of adjacent buildings in NS and EW directions. The analysis results of this study show that placing fluid viscous dampers at selected floors will result in a more efficient structural system to mitigate the earthquake's effects.

Finally, while finding the objective function of the optimal arrangement for each specific number of dampers, the effectiveness of responses to the number of placed dampers is examined. The numerical results prove that increasing the number of dampers does not necessarily improve the efficiency of the system; hence, reducing the cost of dampers. In fact, increasing the number of dampers can even result in increasing the inter-storey drift. The most likely reason for this assumption has been widely used in many studies. The results show that it is not valid and can lead to very inefficient damping values. An interesting research direction to extend the procedures in this study involves optimal passive, active and semi-active control devices for adjacent buildings, considering the SSI system by using multi-objective genetic algorithm in conjunction with fuzzy logic control. The research is utilised for exploring the design space to find an optimal set of solutions.

References

Abdel Raheem, S., Hayashikawa, T. and Dorka, U. 2011. Ground motion spatial variability effects on seismic response control of cable-stayed bridges. Earthquake Engineering and Engineering Vibration, 10(1): 37–49.

Agarwal, V.K., Niedzwecki, J.M. and van de Lindt, J.W. 2007. Earthquake-induced pounding in friction varying base-isolated buildings. Engineering Structures, 29(11): 2825–2832.

Ahlawat, A.S. and Ramaswamy, A. 2002. Multi-objective optimal design of flc driven hybrid mass damper for seismically excited structures. Earthquake Engineering & Structural Dynamics, 31(7): 1459–1479.

Ahlawat, A.S. and Ramaswamy, A. 2003. Multi-objective optimal absorber system for torsionally coupled seismically excited structures. Engineering Structures, 25: 941–950.

Aldemir, U. 2010. A simple active control algorithm for earthquake-excited structures. Computer-Aided Civil and Infrastructure Engineering, 25(3): 218–225.

Ali, S.F. and Ramaswamy, A. 2009. Optimal fuzzy logic control for MDOF structural systems using evolutionary algorithms. Engineering Applications of Artificial Intelligence, 22(3): 407–419.

Anagnostopoulos, S.A. and Spiliopoulos, K.V. 1992. An investigation of earthquake-induced pounding between adjacent buildings. Earthquake Engineering and Structural Dynamics, 21: 302–289.

Arfiadi, Y. 2000. Optimal passive and active control mechanisms for seismically-excited buildings. Faculty of Engineering, University of Wollongong, Wollongong, Doctor of Philosophy thesis.

Arfiadi, Y. and Hadi, M.N.S. 2000. Passive and active control of three-dimensional buildings. Earthquake Engineering & Structural Dynamics, 29(3): 377–396.

Arfiadi, Y. and Hadi, M.N.S. 2001. Optimal direct (static) output feedback controller using real coded genetic algorithms. Computers & Structures, 79(17): 1625–1634.

Arfiadi, Y. and Hadi, M.N.S. 2006. Continuous bounded controllers for active control of structures. Computers & Structures, 84(12): 798–807.

Arfiadi, Y. and Hadi, M.N.S. 2011. Optimum placement and properties of tuned mass dampers using hybrid genetic algorithms. Int. J. Optim. Civil Eng., 1: 167–187.

Balendra, T., Chan Weng, T. and Seng-Lip, L. 1982. Modal damping for torsionally coupled buildings on elastic foundation. Earthquake Engineering & Structural Dynamics, 10(5): 735–756.

Balendra, T. 1983. A simplified model for lateral load analysis of asymmetrical buildings. Engineering Structures, 5(3): 154–162.

Balendra, T., Chan Weng, T. and Seng-Lip, L. 1983. Vibration of asymmetrical building-foundation systems. Journal of Engineering Mechanics, 109(2): 430–449.

Balendra, T. and Koh, C.G. 1991. Vibrations of non-linear asymmetric buildings on a flexible foundation. Journal of Sound and Vibration, 149(3): 361–374.

Battaini, M., Casciati, F. and Faravelli, L. 1998. Fuzzy control of structural vibration. An active mass system driven by a fuzzy controller. Earthquake Engineering & Structural Dynamics, 27(11): 1267–1276.

Bharti, S. and Shrimali, M. 2007. Seismic performance of connected buildings with MR dampers. In: Proceedings of the 8th Pacific Conference on Earthquake Engineering, 3–5 December Singapore: Nangyang Technological University 1–12.

Bharti, S.D., Dumne, S.M. and Shrimali, M.K. 2010. Seismic response analysis of adjacent buildings connected with MR dampers. Engineering Structures, 32(8): 2122–2133.

Bhaskararao, A.V. and Jangid, R.S. 2006a. Seismic analysis of structures connected with friction dampers. Engineering Structures, 28(5): 690–703.

Bhaskararao, A.V. and Jangid, R.S. 2006b. Seismic response of adjacent buildings connected with friction dampers. Bulletin of Earthquake Engineering, 4(1): 43–64.

Bhasker Rao, P. and Jangid, R.S. 2001. Performance of sliding systems under near-fault motions. Nuclear Engineering and Design, 203(2-3): 259–272.

Bigdeli, K., Hare, W. and Tesfamariam, S. 2012. Configuration optimization of dampers for adjacent buildings under seismic excitations. Engineering Optimization, 1–19.

Bitaraf, M., Ozbulut, O.E., Hurlebaus, S. and Barroso, L. 2010. Application of semi-active control strategies for seismic protection of buildings with MR dampers. Engineering Structures, 32(10): 3040–3047.

Bitaraf, M., Hurlebaus, S. and Barroso, L.R. 2012. Active and semi-active adaptive control for undamaged and damaged building structures under seismic load. Computer-aided Civil and Infrastructure Engineering, 27(1): 48–64.

BSSC. 1997. NEHRP guidelines for the seismic rehabilitation of buildings, in FEMA Publication No. 273. Building Seismic Safety Council: Washington, D.C.

Catal, H.H. 2002. Matrix Methods in Structural and Dynamic Analysis. Izmir, Turkey: Department of Civil Engineering, Dokuz Eylul University.

Chang, C. and Zhou, L. 2002. Neural network emulation of inverse dynamics for a magnetorheological damper. Journal of Structural Engineering, 128(2): 231–239.

Cheng, F.Y., Jiang, H. and Lou, K. 2008. Smart Structures: Innovative Systems for Seismic Response Control. New York: CRC Press/Taylor & Francis Group.

Cheng, H., Zhu, W.Q. and Ying, Z.G. 2006. Stochastic optimal semi-active control of hysteretic systems by using a magneto-rheological damper. Smart Materials and Structures, 15(3): 711.

Chopra, A.K. 1995. Dynamics of Structures. Theory and Applications to Earthquake Engineering. New Jersey: Englewood Cliffs: Prentice-Hall.

Chopra, A.K. and Goel, R.K. 2004. A modal pushover analysis procedure to estimate seismic demands for unsymmetric-plan buildings. Earthquake Engineering & Structural Dynamics, 33(8): 903–927.

Christenson, R.E., Spencer, B.F. and Johnson, E.A. 1999. Coupled building control using active and smart damping strategies. *In*: Proc., 5th Int. Conf. on Application of Artificial Intelligence to Civil Engineering and Structural Engineering, Edinburgh, Scotland: Civil-Comp Press, 187–195.

Clough, R.W. and Penzien, J. 1993. Dynamics of Structures; 2nd ed., New York: McGraw-Hill Book Co., Inc.

Constantinou, M., Mokha, A. and Reinhorn, A. 1990. Teflon bearings in base isolation ii: Modelling. Journal of Structural Engineering, 116(2): 455–474.

Council, I.C. 2000. International Building Code 2000: International Code Council.

Dargush, G.F. and Sant, R.S. 2005. Evolutionary aseismic design and retrofit of structures with passive energy dissipation. Earthquake Engineering & Structural Dynamics, 34(13): 1601–1626.

Deb, K. and Agarwal, R.B. 1995. Simulated binary crossover for continuous search space. Complex Systems, 9: 115–148.

Deb, K., Pratap, A., Agarwal, S. and Meyarivan, T. 2002. A fast and elitist multi-objective genetic algorithm: Nsga-ii. Evolutionary Computation, IEEE Transactions, 6(2): 182–197.

DesRoches, R. and Muthukumar, S. 2002. Effect of pounding and restrainers on seismic response of multiple-frame bridges. Journal of Structural Engineering, 128(7): 860–869.

Dumne, S.M. and Shrimali, M.K. 2007. Earthquake performance of isolated buildings connected with MR dampers. *In*: 8th Pacific Conference on Earthquake Engineering, 5–7 December Singapore-244.

Dyke, S.J. 1996. Acceleration feedback control strategies for active and semi-active control systems. Modelling, Algorithm Development, and Experimental Verification. Department of Civil Engineering and Geological Sciences, University of Notre Dame, Notre Dame, Indiana, Doctor of Philosophy.

Dyke, S.J.,Spencer, J.B.F., Sain, M.K. and Carlson, J.D. 1996a. Modelling and control of magnetorheological dampers for seismic response reduction. Smart Materials and Structures, 5(5): 565.

Dyke, S.J., Spencer, J.B.F., Quast, P., Sain, M.K., Kaspari Jr., D.C. and Soong, T.T. 1996b. Acceleration feedback control of MDOF structures. Journal of Engineering Mechanics, 122(9): 907–918.

Dyke, S.J., Spencer, J.B.F., Sain, M.K. and Carlson, J.D. 1998. An experimental study of MR dampers for seismic protection. Smart Materials and Structures, 7(5): 693.

El-Alfy, E.S.M. 2010. Flow-based path selection for internet traffic engineering with nsga-ii. *In*: Telecommunications (ICT), 2010 IEEE 17th International Conference on, 4–7 April 2010, 621–627.

Fallahpour, M.B., Hemmati, K.D. and Pourmohammad, A. 2012. Optimization of a lna using genetic algorithm. Electrical and Electronic Engineering, 2(2): 38–42.

Goldberg, D.E. 1989. Genetic Algorithms in Search, Optimization and Machine Learning. MA: Addison-Wesley, Reading.

Gupta, V.K. and Trifunac, M.D. 1991. Seismic response of multistoried buildings including the effects of soil-structure interaction. Soil Dynamics and Earthquake Engineering, 10(8): 414–422.

Hadi, M.N.S. and Arfiadi, Y. 1998. Optimum design of absorber for MDOF structures. Journal of Structural Engineering, 124(11): 1272–1280.

Hadi, M.N.S. and Uz, M.E. 2009. Improving the dynamic behaviour of adjacent buildings by connecting them with fluid viscous dampers. 2nd International Conference on Computational Methods in Structural Dynamics and Earthquake Engineering, COMPDYN 2009, June 22–24 Island of Rhodes, Greece: Institute of Structural Analysis & Seismic Research National Technical University of Athens, 280.

Hadi, M.N. and Uz, M. 2010a. Base-isolated adjacent buildings considering the effect of pounding and impact due to earthquakes. International Congress on Advances in Civil Engineering, Trabzon, Turkey: Eser Ofset Matbaacilik, pp. 227–227.

Hadi, M.N.S. and Uz, M.E. 2010b. Inelastic base-isolated adjacent buildings under earthquake excitation with the effect of pounding. The 5th Civil Engineering Conference in the Asian Region and Australasian Structural Engineering Conference 2010 CECAR 5/ASEC 2010, 8–12 Aug Sydney, Australia, 155–201.

Hejal, R. and Chopra, A.K. 1989. Earthquake analysis of a class of torsionally-coupled buildings. Earthquake Engineering & Structural Dynamics, 18(3): 305–323.

Herrera, F., Lozano, M. and Verdegay, J.L. 1998. Tackling real-coded genetic algorithms: Operators and tools for behavioral analysis. Artif. Intell. Rev., 12(4): 265–319.

Holland, J.H. 1975. Adaptation in Natural and Artificial Systems: The University of Michigan Press, Ann Arbor.

Holland, J.H. 1992. Adaptation in Natural and Artificial Systems: MIT Press, Mass.

Hou, C.Y. 2008. Fluid dynamics and behaviour of non-linear viscous fluid dampers. Journal of Structural Engineering, 134(1): 56–63.

Housner, G.W., Soong, T.T. and Masri, S.F. 1994. Second generation of active structural control in civil engineering. Proceedings of the First World Conference on Structural Control, Pasadena, California.

International Building Code (IBC). 2003. International Code Council, USA.

Iwanami, K., Suzuki, K. and Seto, K. 1996. Vibration control method for parallel structures connected by damper and spring. JSME International Journal, Series C 39: 714–720.

Jankowski, R., Wilde, K. and Fujino, Y. 1998. Pounding of superstructure segments in isolated elevated bridge during earthquakes. Earthquake Engineering & Structural Dynamics, 27(5): 487–502.

Jankowski, R. 2006. Pounding force response spectrum under earthquake excitation. Engineering Structures, 28(8): 1149–1161.

Jankowski, R. 2008. Earthquake-induced pounding between equal height buildings with substantially different dynamic properties. Engineering Structures, 30(10): 2818–2829.

Jankowski, R. 2010. Experimental study on earthquake-induced pounding between structural elements made of different building materials. Earthquake Engineering & Structural Dynamics, 39(3): 343–354.

Jansen, L.M. and Dyke, S.J. 1999. Investigation of non-linear control strategies for the implementation of multiple magnetorheological dampers. *In*: Proceedings of the Engineering Mechanics Conference, June 13–16 Baltimore, Maryland: ASCE.

Jansen, L.M. and Dyke, S.J. 2000. Semi-active control strategies for MR dampers: Comparative study. Journal of Engineering Mechanics, 126(8): 795–803.

Jui-Liang, L. and Keh-Chyuan, T. 2007. Simplified seismic analysis of one-way asymmetric elastic systems with supplemental damping. Earthquake Engineering & Structural Dynamics, 36(6): 783–800.

Jui-Liang, L., Keh-Chyuan, T. and Eduardo, M. 2009. Seismic history analysis of asymmetric buildings with soil-structure interaction. Journal of Structural Engineering, 135(2): 101–112.

Jung, H., Spencer, B. and Lee, I. 2003. Control of seismically excited cable-stayed bridge employing magnetorheological fluid dampers. Journal of Structural Engineering, 129(7): 873–883.

Jung, H.-J., Choi, K.-M., Spencer, B.F. and Lee, I.-W. 2006. Application of some semi-active control algorithms to a smart base-isolated building employing MR dampers. Structural Control and Health Monitoring, 13(2-3): 693–704.

Kageyama, M., Yoshida, O. and Yasui, Y. 1994. A study on optimum damping systems for connected double frame structures. Proceeding of the First Conference on Structural Control, 1: 4–32.

Kalasar, H.E., Shayeghi, A. and Shayeghi, H. 2009. Seismic control of tall buildings using a new optimum controller based on GA. International Journal of Applied Science, Engineering and Technology, 5(2): 85.

Kan, C.L. and Chopra, A.K. 1976. Coupled lateral torsional response of buildings to ground motion, Report No. EERC 76-13, Earthquake Engineering Research Center, University of California, Brekeley, Calif.

Kasai, K., Jagiasi, A.R. and Jeng, V. 1996. Inelastic vibration phase theory for seismic pounding mitigation. Journal of Structural Engineering, 122(10): 1136–1146.

Kim, H.-S. and Roschke, P.N. 2006. Fuzzy control of base-isolation system using multi-objective genetic algorithm. Computer-aided Civil and Infrastructure Engineering, 21(6): 436–449.

Kim, H.-S. and Kang, J.-W. 2012. Semi-active fuzzy control of a wind-excited tall building using multi-objective genetic algorithm. Engineering Structures, 41(0): 242–257.

Kim, J., Ryu, J. and Chung, L. 2006. Seismic performance of structures connected by viscoelastic dampers. Engineering Structures, 28(2): 183–195.

Kim, Y.-J. and Ghaboussi, J. 2001. Direct use of design criteria in genetic algorithm-based controller optimization. Earthquake Engineering & Structural Dynamics, 30(9): 1261–1278.

Kirk, D.E. 1970. Optimal Control Theory—An Introduction: Englewood Cliffs, NJ: Prentice-Hall.

Kitagawa, Y. and Midorikawa, M. 1998. Seismic isolation and passive response-control buildings in japan. Smart Materials and Structures, 7(5): 581.

Klein, R.G. and Healy, M.D. 1985. Semiactive control of wind induced oscillations in structures. Proceedings of the 2nd International Conference on Structural Control, University of Waterloo, Ontario, Canada, 187–195.

Kurihara, K., Haramoto, H. and Seto, K. 1997. Vibration control of flexible structure arranged in parallel in response to large earthquakes. *In*: Proceedings of the Asia-Pacific Vibration Conference 97, Kyongju, Korea, 1205–1210.

Lavan, O. and Dargush, G.F. 2009. Multi-objective evolutionary seismic design with passive energy dissipation systems. Journal of Earthquake Engineering, 13(6): 758–790.

Levine, W.S. and Athans, M. 1970. On the determination of the optimal constant output feedback gains for linear multi-variable systems. IEEE Transactions on Automatic Control, 15(1): 44–48.

Lewis, F. and Syrmos, V. 1995. Optimal Control. New York: John Wiley & Sons.

Lin, J.-L. and Tsai, K.-C. 2007. Simplified seismic analysis of asymmetric building systems. Earthquake Engineering & Structural Dynamics, 36(4): 459–479.

Lin, J.H. and Weng, C.C. 2001. Probability seismic pounding of adjacent buildings. Earthquake Engineering and Structural Dynamics, 30: 1539–1557.

Lopez-Garcia, D. and Soong, T.T. 2009. Assessment of the separation necessary to prevent seismic pounding between linear structural systems. Probabilistic Engineering Mechanics, 24(2): 210–223.

270

Lu, L. 2001. Extended LQG methodology for active structural controller design. National Central University, China, Ph.D. Dissertation.

Lublin, L., Grocott, S. and Athans, M. 1996. H_2 (LQG) and H_∞ control. *In*: The Control Handbook. Boca Raton: CRC Press. pp. 635–650.

Luco, J.E. and Wong, H.L. 1994. Control of the seismic response of adjacent structures. Proceedings of the First World Conference on Structural Control, Los Angeles CA, 21–30.

Luco, J.E. and De Barros, F.C.P. 1998. Optimal damping between two adjacent elastic structures. Earthquake Engineering & Structural Dynamics, 27(7): 649–659.

Mahmoud, S. and Jankowski, R. 2009. Elastic and inelastic multi-storey buildings under earthquake excitation with the effect of pounding. Journal of Applied Sciences, 9(18): 3250–3262.

Maison, B.F. and Kasai, K. 1992. Dynamics of pounding when two buildings collide. Earthquake Engineering & Structural Dynamics, 21(9): 771–786.

MATLAB, R. 2011b. The Math Works, Inc.: Natick, MA.

Matsagar, V.A. and Jangid, R.S. 2003. Seismic response of base-isolated structures during impact with adjacent structures. Engineering Structures, 25(10): 1311–1323.

Meirovitch, L. 1992. Dynamics and control of structures. Singapore: John Wiley & Sons.

Michalewicz, Z. 1996. Genetic Algorithms + Data Structures = Evolution Programs: Springer, Berlin.

Mitchell, M. 1998. An Introduction to Genetic Algorithms. Cambridge, MA: A Bradford Book, The MIT Press.

Mokha, A., Constantinou, M. and Reinhorn, A. 1990. Teflon bearings in base isolation: Testing. Journal of Structural Engineering, 116(2): 438–454.

Motra, G.B., Mallik, W. and Chandiramani, N.K. 2011. Semi-active vibration control of connected buildings using magnetorheological dampers. Journal of Intelligent Material Systems and Structures, 22(16): 1811–1827.

Ng, C.-L. and Xu, Y.-L. 2006. Seismic response control of a building complex utilising passive friction damper: Experimental investigation. Earthquake Engineering & Structural Dynamics, 35(6): 657–677.

Nguyen, H.T., Sugeno, M., Tong, R.M. and Yager, R.R. 1995. Theoretical Aspects of Fuzzy Control. New York: Wiley.

Ni, Y.Q., Ko, J.M. and Ying, Z.G. 2001. Random seismic response analysis of adjacent buildings coupled with non-linear hysteretic dampers. Journal of Sound and Vibration, 246(3): 403–417.

NOAA. 2008. Natural hazards, significant earthquake database, NOAA. National Geophysical Data Center, Boulder, Colorado, http://www.ngdc.noaa.gov/.

Novak, M. and Hifnawy, L.E. 1983a. Damping of structures due to soil-structure interaction. Journal of Wind Engineering and Industrial Aerodynamics, 11(1-3): 295–306.

Novak, M. and Hifnawy, L.E. 1983b. Effect of soil-structure interaction on damping of structures. Earthquake Engineering & Structural Dynamics, 11(5): 595–621.

Ok, S.-Y., Kim, D.-S., Park, K.-S. and Koh, H.-M. 2007. Semi-active fuzzy control of cable-stayed bridges using magneto-rheological dampers. Engineering Structures, 29(5): 776–788.

Ok, S.-Y., Song, J. and Park, K.-S. 2008. Optimal design of hysteretic dampers connecting adjacent structures using multi-objective genetic algorithm and stochastic linearization method. Engineering Structures, 30(5): 1240–1249.

Papadrakakis, M., Apostolopoulou, C., Zacharopoulos, A. and Bitzarakis, S. 1996. Three-dimensional simulation of structural pounding during earthquakes. Journal of Engineering Mechanics, 122(5): 423–431.

Park, K.-S., Koh, H.-M. and Ok, S.-Y. 2002. Active control of earthquake excited structures using fuzzy supervisory technique. Adv. Eng. Softw., 33(11-12): 761–768.

Park, K.-S., Koh, H.-M., Ok, S.-Y. and Seo, C.-W. 2005. Fuzzy supervisory control of earthquake-excited cable-stayed bridges. Engineering Structures, 27(7): 1086–1100.

Patel, C.C. and Jangid, R.S. 2009. Seismic response of dynamically similar adjacent structures connected with viscous dampers. The IES Journal Part A: Civil & Structural Engineering, 3(1): 1–13.

271

Penzien, J. 1997. Evaluation of building separation distance required to prevent pounding during strong earthquakes. Earthquake Engineering & Structural Dynamics, 26(8): 849–858.

Pourzeynali, S., Lavasani, H.H. and Modarayi, A.H. 2007. Active control of high rise building structures using fuzzy logic and genetic algorithms. Engineering Structures, 29(3): 346–357.

Qi, X.X. and Chang, K.L. 1995. Study of application of viscous dampers in seismic joints. Proceeding of the International Conference on Structural Dynamics, Hong Kong Vibration, Noise and Control.

Rabinow, J. 1948. The magnetic fluid clutch. American Institute of Electrical Engineers, Transactions, 67(2): 1308–1315.

Richard, E.C., Spencer, B.F. Jr., Erik, A.J. and Seto, K. 2006. Coupled building control considering the effects of building/connector configuration. Journal of Structural Engineering, 132(6): 853–863.

Richart, F.E., Hall, J.R. and Woods, R.D. 1970. Vibrations of Soils and Foundations. N.J: Prentice-Hall, Englewood Cliffs.

Robert, J. 2009. Experimental study on earthquake-induced pounding between structural elements made of different building materials. Earthquake Engineering & Structural Dynamics, 39(3): 343–354.

Sadek, F., Mohraz, B., Taylor, A.W. and Chung, R.M. 1997. A method of estimating the parameters of tuned mass dampers for seismic applications. Earthquake Engineering & Structural Dynamics, 26(6): 617–635.

Schurter, K.C. and Roschke, P.N. 2001. Neuro-fuzzy control of structures using magnetorheological dampers. *In*: American Control Conference, 2001. Proceedings of the 2001, 2: 1097–1102.

Seleemah, A.A. and Constantinou, M.C. 1997. Investigation of seismic response of buildings with linear and non-linear fluid viscous dampers, NCEER-97-0004, National Center for Earthquake Research Rep., State Univ. of New York at Buffalo, Buffalo, New York.

Seto, K. and Mitsuta, S. 1992. Active vibration control of structures arranged in parallel. *In*: Proceedings of the First International Conference on Motion and Vibration Control, Japan 146–151.

Seto, K. 1994. Vibration control method for flexible structures arranged in parallel. Proc. First World Conference on Structural Control, Los Angeles, FP3: 62–71.

Shook, D.A., Roschke, P.N. and Ozbulut, O.E. 2008. Superelastic semi-active damping of a base-isolated structure. Structural Control and Health Monitoring, 15(5): 746–768.

Singh, M.P., Matheu, E.E. and Suarez, L.E. 1997. Active and semi-active control of structures under seismic excitation. Earthquake Engineering & Structural Dynamics, 26(2): 193–213.

Sivakumaran, K.S., Lin, M.-S. and Karasudhi, P. 1992. Seismic analysis of asymmetric building-foundation systems. Computers & Structures, 43(6): 1091–1103.

Sivakumaran, K.S. and Balendra, T. 1994. Seismic analysis of asymmetric multistorey buildings including foundation interaction and p-[delta] effects. Engineering Structures, 16(8): 609–624.

Smith, R.E. 1993. Adaptively resizing populations: An algorithm and analysis. *In*: Proc. of The 5th International Conference on Genetic Algorithms, San Francisco, CA, USA: Morgan Kaufmann Publishers Inc.

Soong, T.T. and Dargush, G.F. 1997. Passive energy dissipation systems in structural engineering, John Wiley & Sons, Chichester, New York.

Spencer, J.B.F., Suhardjo, J. and Sain, M.K. 1994. Frequency domain optimal control strategies for aseismic protection. Journal of Engineering Mechanics, 120(1): 135–158.

Spencer Jr., B.F., Dyke, S.J., Sain, M.K. and Carlson, J.D. 1997. Phenomenological model for magnetorheological dampers. Journal of Engineering Mechanics, 123(3): 230–238.

Spencer, B.F., Dyke, S.J. and Deoskar, H.S. 1998. Benchmark problems in structural control: Part I—active mass driver system. Earthquake Engineering & Structural Dynamics, 27(11): 1127–1139.

Spyrakos, C.C., Koutromanos, I.A. and Maniatakis, C.A. 2009. Seismic response of base-isolated buildings including soil-structure interaction. Soil Dynamics and Earthquake Engineering, 29(4): 658–668.

Srinivas, N. and Deb, K. 1994. Multi-objective optimization using nondominated sorting in genetic algorithms. Evol. Comput., 2(3): 221–248.

Stanway, R., Sproston, J.L. and Stevens, N.G. 1987. Non-linear modelling of an electro-rheological vibration damper. Journal of Electrostatics, 20(2): 167–184.

Sugino, S., Sakai, D., Kundu, S. and Seto, K. 1999. Vibration control of parallel structures connected with passive devices designed by GA. Proceedings of the Second world Conference on Structural Control: Chichester: John Wiley & Sons, pp. 329–337.

Sun, H., Lus, H. and Betti, R. 2013. Identification of structural models using a modified Artificial Bee Colony algorithm. Computer and Structures, 116(0): 59–74.

Symans, M.D. and Constantinou, M.C. 1999. Semi-active control systems for seismic protection of structures: A state-of-the-art review. Engineering Structures, 21(6): 469–487.

Symans, M.D. and Kelly, S.W. 1999. Fuzzy logic control of bridge structures using intelligent semi-active seismic isolation systems. Earthquake Engineering & Structural Dynamics, 28(1): 37–60.

Taylor, D.P. and Constantinou, M.C. 1998. Development and testing of an improved fluid damper configuration for structures having high rigidity. Proceeding of the 69th Shock and Vibration Symposium.

Tezcan, S.S. and Uluca, O. 2003. Reduction of earthquake response of plane frame buildings by viscoelastic dampers. Engineering Structures, 25(14): 1755–1761.

Thambirajah, B., Chan Weng, T. and Seng-Lip, L. 1982. Modal damping for torsionally coupled buildings on elastic foundation. Earthquake Engineering & Structural Dynamics, 10(5): 735–756.

Thambirajah, B., Chan Weng, T. and Seng-Lip, L. 1983. Vibration of asymmetrical building-foundation systems. Journal of Engineering Mechanics. 109(2): 430–449.

Tsai, N.C., Niehoff, D., Swatta, M. and Hadjian, A.H. 1975. The use of frequency-independent soil-structure interaction parameters. Nuclear Engineering and Design, 31(2): 168–183.

Uz, M.E. 2009. Improving the dynamic behaviour of adjacent buildings by connecting them with fluid viscous dampers. School of Civil, Mining Environmental, Engineering, University of Wollongong, Wollongong, ME (Res) Thesis.

Uz, M.E. and Hadi, M.N.S. 2009. Dynamic analyses of adjacent buildings connected by fluid viscous dampers. Seventh World Conference on Earthquake Resistant Engineering Structures ERES VII, Limassol, Cyprus: Wessex Institute of Technology, pp. 139–150.

Uz, M.E. and Hadi, M.N.S. 2010. Investigating the effect of pounding for base isolated adjacent buildings under ground motion. 21st Australasian Conference on the Mechanics of Structures and Materials Melbourne, 7–10 December Victoria, Australia.

Uz, M.E. 2013. Optimum design of semi-active dampers between adjacent buildings of different sizes subjected to seismic loading including soil–structure interaction. School of Civil, Mining and Environmental Eng, University of Wollongong, Wollongong, Doctor of Philosophy Thesis.

Valles, R.E. and Reinhorn, A.M. 1997. Evaluation, prevention and mitigation of pounding effects in buildings structures, National Center of Earthquake Engineering Research Technical Report.

Warnotte, V., Stoica, D., Majewski, S. and Voiculescu, M. 2007. State of the art in the pounding mitigation techniques. Structural Mechanics, Intersections/Intersectii, 4, No. 3.

Wen, Y.K. 1976. Method for random vibration of hysteretic systems. Journal of the Engineering Mechanics Division, Am. Soc. Civ. Eng., 102(2): 249–263.

Westermo, B. 1989. The dynamics of interstructural connection to prevent pounding. Earthquake Engineering and Structural Dynamics, 18: 687–699.

Wriggers, P. 2006a. Analysis and simulation of contact problems. Lecture notes in applied and computational mechanics. Vol. 27. Berlin; Heidelberg; New York: Springer.

Wriggers, P. 2006b. Computational contact mechanics with 12 tables. Berlin; Heidelberg; New York: Springer.

Wu, W.H., Wang, J.F. and Lin, C.C. 2001. Systematic assessment of irregular building–soil interaction using efficient modal analysis. Earthquake Engineering & Structural Dynamics, 30(4): 573–594.

Xu, Y.L., He, Q. and Ko, J.M. 1999. Dynamic response of damper-connected adjacent buildings under earthquake excitation. Engineering Structures, 21(2): 135–148.

Xu, Y.L. and Zhang, W.S. 2002. Closed-form solution for seismic response of adjacent buildings with linear quadratic gaussian controllers. Earthquake Engineering & Structural Dynamics, 31(2): 235–259.

Yamada, Y., Ikawa, N., Yokoyama, H. and Tachibana, E. 1994. Active control of structures using the joining member with negative stiffness. Proceedings of the First World Conference on Structural Control, Los Angeles: CA, 41–49.

Yan, G. and Zhou, L.L. 2006. Integrated fuzzy logic and genetic algorithms for multi-objective control of structures using MR dampers. Journal of Sound and Vibration, 296(1-2): 368–382.

Yao, J.T.P. 1972. Concept of structural control. ASCE J. Struct. Div., 98: 1567–1574.

Yang, G., Spencer Jr., B.F., Carlson, J.D. and Sain, M.K. 2002. Large-scale MR fluid dampers: Modeling and dynamic performance considerations. Engineering Structures, 24(3): 309–323.

Yang, Z., Xu, Y.L. and Lu, X.L. 2003. Experimental seismic study of adjacent buildings with fluid dampers. Journal of Structural Engineering, 129(2): 197–205.

Ying, Z.G., Ni, Y.Q. and Ko, J.M. 2003. Stochastic optimal coupling-control of adjacent building structures. Computers & Structures, 81(30-31): 2775–2787.

Yoshida, O., Dyke, S.J., Giacosa, L.M. and Truman, K.Z. 2003. Experimental verification of torsional response control of asymmetric buildings using MR dampers. Earthquake Engineering & Structural Dynamics, 32(13): 2085–2105.

Yoshida, O. and Dyke, S.J. 2004. Seismic control of a non-linear benchmark building using smart dampers. Journal of Engineering Mechanics, 130(4): 386–392.

Zadeh, L.A. 1965. Fuzzy sets. Information Control 8: 338–53.

Zhang, W.S. and Xu, Y.L. 1999. Dynamic characteristics and seismic response of adjacent buildings linked by discrete dampers. Earthquake Engineering & Structural Dynamics, 28(10): 1163–1185.

Zhang, W.S. and Xu, Y.L. 2000. Vibration analysis of two buildings linked by maxwell model-defined fluid dampers. Journal of Sound and Vibration, 233(5): 775–796.

Zhou, L., Chang, C. and Wang, L. 2003. Adaptive fuzzy control for nonlinear building-magnetorheological damper system. Journal of Structural Engineering, 129(7): 905–913.

Zhu, H.P. and Xu, Y.L. 2005. Optimum parameters of maxwell model-defined dampers used to link adjacent structures. Journal of Sound and Vibration, 279(1-2): 253–274.

Zhu, H.P., Ge, D.D. and Huang, X. 2011. Optimum connecting dampers to reduce the seismic responses of parallel structures. Journal of Sound and Vibration, 330(9): 1931–1949.

Appendices
Data of the Buildings given in Chapter 7

Appendix A-1: The Design Data

Table A1 The Design Data for All Examples.

Live load	: 4.0 kN/m^2 at typical floor
	: 1.5 kN/m^2 on roof
Floor finish	: 1.0 kN/m^2
Water proofing	: 2.0 kN/m^2 on roof
Terrace finish	: 1.0 kN/m^2
Location	: Structures in New Zealand
Importance Levels	: 2 (Buildings not included in importance levels 1, 3 or 4) (AS/NZS 1170.0-2002)
Earthquake load	: As per AS/NZS 1170.4-2002 Time history analyses
Depth of foundation below ground	: 2.5 m
Type of soil	: Type II, Medium as per AS/NZS 1170.0-2002
Allowable bearing pressure	: 200 kN/m^2
Walls	: 230 mm thick brick masonry walls only at periphery.

Appendix A-2: Material Properties

Table A2 Material Properties for Concrete and Steel.

Concrete
All components unless specified in design: M 30 grade
For M 30, fck = 30 MPa
$E_c = 5000 \times \sqrt{f_{ck}} = 27386$ N/mm²
Where fck is characteristic strength of concrete and Ec stands for elastic modulus of concrete at 28 days
Concrete density = 25 kN/m³
Wall Density = 19 kN/m³
Plaster Density = 20 kN/m³
Parapet = 4.85 kN/m²
Heights of materials:
Parapet = 1 m
Slab = 0.12 m
External Wall = 0.23 m
Internal Wall = 0.1 m
Plaster Thick = 0.012 m
Steel
HYSD reinforcement of grade Fe 415 confirming to IS:1786 is used throughout

Appendix A-3: Gravity Load Calculations for Examples

Table A3 Unit Load Calculations for Examples 1 and 2(a).

Assumed sizes of beams and column sections are:
Columns: 300 × 600 mm² in Building *A*
Area, A = 0.18 m²
Columns: 300 × 500 mm² in Building *B*
Area, A = 0.15 m²
Beams: 250 × 600 mm² in Building *A*
Area, A = 0.15 m²
Beams: 250 × 500 mm² in Building *B*
Area, A = 0.125 m²

Table A4 Member Self-Weights for Examples 1 and 2(a).

Member Self-weights: Building A		Member Self-weights: Building B	
Columns =	4.5 kN/m	Columns =	3.75 kN/m
Beams =	3.75 kN/m	Beams =	3.125 kN/m
Slab (120 mm thick) =	3 kN/m^2	Slab (120 mm thick) =	3 kN/m^2
External brick wall (230 mm thick) =	4.85 kN/m^2	External brick wall (230 mm thick) =	4.85 kN/m^2
Internal brick wall (100 mm thick) =	2.38 kN/m^2	Internal brick wall (100 mm thick) =	2.38 kN/m^2
Roof Parapet (height 1.0 m) =	4.85 kN/m	Roof Parapet (height 1.0 m) =	4.85 kN/m

Table A5 Unit Load Calculations for Examples 3 and 4(a).

> Assumed sizes of beams and column sections are:
> Columns: 300×700 mm^2 in Building A
> Area, A = 0.21 m^2
> Columns: 300×600 mm^2 in Building B
> Area, A = 0.18 m^2
> Beams: 300×600 mm^2 in Building A
> Area, A = 0.18 m^2
> Beams: 300×500 mm^2 in Building B
> Area, A = 0.15 m^2

Table A6 Member Self-weights for Examples 3 and 4(a).

Member Self-weights: Building A		Member Self-weights: Building B	
Columns =	7 kN/m	Columns =	4.5 kN/m
Beams =	4.5 kN/m	Beams =	3.75 kN/m
Slab (120 mm thick) =	3 kN/m^2	Slab (120 mm thick) =	3 kN/m^2
External brick wall (230 mm thick) =	4.85 kN/m^2	External brick wall (230 mm thick) =	4.85 kN/m^2
Internal brick wall (100 mm thick) =	2.38 kN/m^2	Internal brick wall (100 mm thick) =	2.38 kN/m^2
Roof Parapet (height 1.0 m) =	4.85 kN/m	Roof Parapet (height 1.0 m) =	4.85 kN/m

Appendix A-4: Slab Load Calculations and Beam-Frame Load Calculations for All Examples in Chapter 7

Table A7 Slab Load Calculations.

Slab load calculations:				
Building *A* and *B*				
	Roof		Floor	
Component (kN/m²)	DL	LL	DL	LL
Self (120 mm thick)	3	0	3	0
Water proofing	2	0	0	0
Floor finish	1	0	1	0
Live load	0	1.5	0	4
Total (kN/m²)	6	1.5	4	4

Table A8 Beam and Frame Load Calculations.

Building *A* and *B*		
Beam and Frame load calculations:		
Roof Level		
Internal Beams	DL	LL
From Slab	18	4.5
Total (kN/m)	18	4.5
External Beams	DL	LL
From Slab	9	2.25
Parapet	4.85	0
Total (kN/m)	13.85	2.25
Floor Level		
Internal Beams	DL	LL
From Slab	12	12
Internal Walls	5.712	0
Total (kN/m)	17.712	12
External Beams	DL	LL
From Slab	6	6
External Walls	11.64	0
Total (kN/m)	17.64	6

Appendix A-5: Seismic Weight of the Floors for Examples

Table A9 Seismic Weight of the Floors for Building *A* in Example 1(a).

Building *A*		
Seismic Weight of the Floors		
Roof Level	DL	LL
From Slab	864	216
Parapet	232.8	0
External Walls	279.36	0
Internal Walls	68.544	0
Main Beams	270	0
Columns	60.75	0
Total	1775.454	216
W(DL+0.3*LL)	1840.254	kN
Floor Level	DL	LL
From Slab	576	576
External Walls	558.72	0
Internal Walls	137.088	0
Main Beams	270	0
Columns	121.5	0
Total	1663.308	576
W(DL+0.3*LL)	1836.108	kN
Story: 5		
Total Structure Weight:	9184.686	kN

Table A10 Seismic Weight of the Floors for Building *B* in Example 1(a).

Building *B*		
Seismic Weight of the Floors		
Roof Level	DL	LL
From Slab	864	216
Parapet	232.8	0
External Walls	279.36	0
Internal Walls	68.544	0
Main Beams	225	0
Columns	50.625	0
Total	1720.329	216
W(DL+0.3*LL)	1785.129	kN
Floor Level	DL	LL
From Slab	576	576
External Walls	558.72	0
Internal Walls	137.088	0
Main Beams	225	0
Columns	101.25	0
Total	1598.058	576
W(DL+0.3*LL)	1770.858	kN
Story: 5		
Total Structure Weight:	8868.561	kN

Table A11 Seismic Weight of the Floors for Building *A* in Example 2(a).

Building *A*		
Seismic Weight of the Floors		
Roof Level	DL	LL
From Slab	864	216
Parapet	232.8	0
External Walls	279.36	0
Internal Walls	68.544	0
Main Beams	270	0
Columns	60.75	0
Total	1775.454	216
W(DL+0.3*LL)	1840.254	kN
Floor Level	DL	LL
From Slab	576	576
External Walls	558.72	0
Internal Walls	137.088	0
Main Beams	270	0
Columns	121.5	0
Total	1663.308	576
W(DL+0.3*LL)	1836.108	kN
Story: 10		
Total Structure Weight:	18365.23	kN

Table A12 Seismic Weight of the Floors for Building *B* in Example 2(a).

Building *B*		
Seismic Weight of the Floors		
Roof Level	DL	LL
From Slab	864	216
Parapet	232.8	0
External Walls	279.36	0
Internal Walls	68.544	0
Main Beams	225	0
Columns	50.625	0
Total	1720.329	216
W(DL+0.3*LL)	1785.129	kN
Floor Level	DL	LL
From Slab	576	576
External Walls	558.72	0
Internal Walls	137.088	0
Main Beams	225	0
Columns	101.25	0
Total	1598.058	576
W(DL+0.3*LL)	1770.858	kN
Story: 5		
Total Structure Weight:	8868.561	kN

Table A13 Seismic Weight of the Floors for Building *A* in Example 3(a).

Building *A*		
Seismic Weight of the Floors		
Roof Level	DL	LL
From Slab	864	216
Parapet	232.8	0
External Walls	279.36	0
Internal Walls	68.544	0
Main Beams	324	0
Columns	94.5	0
Total	1863.204	216
W(DL+0.3*LL)	1928.004	kN
Floor Level	DL	LL
From Slab	576	576
External Walls	558.72	0
Internal Walls	137.088	0
Main Beams	324	0
Columns	189	0
Total	1784.808	576
W(DL+0.3*LL)	1957.608	kN
Story: 20		
Total Structure Weight:	39122.56	kN

Table A14 Seismic Weight of the Floors for Building *B* in Example 3(a).

Building *B*		
Seismic Weight of the Floors		
Roof Level	DL	LL
From Slab	864	216
Parapet	232.8	0
External Walls	279.36	0
Internal Walls	68.544	0
Main Beams	270	0
Columns	60.75	0
Total	1775.454	216
W(DL+0.3*LL)	1840.254	kN
Floor Level	DL	LL
From Slab	576	576
External Walls	558.72	0
Internal Walls	137.088	0
Main Beams	270	0
Columns	121.5	0
Total	1663.308	576
W(DL+0.3*LL)	1836.108	kN
Story: 10		
Total Structure Weight:	18365.226	kN

Table A15 Seismic Weight of the Floors for Building *A* in Example 4(a).

Building *A*		
Seismic Weight of the Floors		
Roof Level	DL	LL
From Slab	864	216
Parapet	232.8	0
External Walls	279.36	0
Internal Walls	68.544	0
Main Beams	324	0
Columns	94.5	0
Total	1863.204	216
W(DL+0.3*LL)	1928.004	kN
Floor Level	DL	LL
From Slab	576	576
External Walls	558.72	0
Internal Walls	137.088	0
Main Beams	324	0
Columns	189	0
Total	1784.808	576
W(DL+0.3*LL)	1957.608	kN
Story: 20		
Total Structure Weight:	39122.56	kN

Table A16 Seismic Weight of the Floors for Building *B* in Example 4(a).

Building *B*		
Seismic Weight of the Floors		
Roof Level	DL	LL
From Slab	864	216
Parapet	232.8	0
External Walls	279.36	0
Internal Walls	68.544	0
Main Beams	270	0
Columns	60.75	0
Total	1775.454	216
W(DL+0.3*LL)	1840.254	kN
Floor Level	DL	LL
From Slab	576	576
External Walls	558.72	0
Internal Walls	137.088	0
Main Beams	270	0
Columns	121.5	0
Total	1663.308	576
W(DL+0.3*LL)	1836.108	kN
Story: 20		
Total Structure Weight:	36726.306	kN

Index

Biography

Dr. Mehmet Eren Uz is lecturer in Department of Civil Engineering at the Adnan Menderes University; formerly, an Associate Research Fellow in Civil Engineering (Structural Engineering) at the University of Wollongong. Prior to this, in 2014, he was awarded the doctorate in philosophy in Structural Engineering from the University of Wollongong with special commendation following the completion of his masters degree at the same university. Earlier, he was a bachelor (Honours) student at the University of Dokuz Eylul.

He has acquired substantial knowledge in advanced optimisation techniques in structural engineering, specifically related to building structures, and since the final year of his PhD study, he has completed three research projects in structural engineering. His research focusses on current ultimate capacity of a structural steel bolted connection, coated products for resilient Australian buildings and existing high steel intensity prefabricated construction systems. He is a member of the Institution of Engineers, Australia (IEAust) and the Australian Steel Institute (ASI).

Muhammad N.S. Hadi is an Associate Professor at the School of Civil, Mining and Environmental Engineering, University of Wollongong, Australia. His research interests are in concrete, FRP and optimisation. He earned his PhD from the University of Leeds, UK and his MSc and BSc from the University of Baghdad, Iraq. He is Fellow of Engineers Australia, Fellow of the American Society of Civil Engineers, Member of the Structural Engineering Institute, Member of the American Concrete Institute, Member of the International Association for Bridge and Structural Engineering, Member of Concrete Institute of Australia, Member of NSW branch, Member of International Institute for FRP in Construction (IIFC) and Life Member Australasian (iron & steel) Slag Association (ASA). He has successfully supervised 15 PhD students and 8 MPhil students. He is currently supervising 20 PhD students. He has attracted $1.8M of research funds alone and with collaboration with my colleagues. He has conducted 19 consultancies with the value of $85k. He has published 118 journal articles, 233 conference articles, 16 reports and three books. He has organised two international conferences and is member of organising committees of several international conferences. He has presented his research findings in over 30 countries.